T0344829

MULTIPHYSICS SIMULATION BY DESIGN FOR ELECTRICAL MACHINES, POWER ELECTRONICS, AND DRIVES

MULTIPHYSICS SIMULATION BY DESIGN FOR ELECTRICAL MACHINES, POWER ELECTRONICS, AND DRIVES

MARIUS ROSU
PING ZHOU
DINGSHENG LIN
DAN IONEL
MIRCEA POPESCU
FREDE BLAABJERG
VANDANA RALLABANDI
DAVID STATON

IEEE
PRESS
SERIES
ON POWER
ENGINEERING

IEEE PRESS

WILEY

Published by John Wiley & Sons, Inc., Hoboken, New Jersey.
Published simultaneously in Canada.

ANSYS and any and all ANSYS, Inc., brand, product, service and feature names, logos and slogans are registered trademarks or trademarks of ANSYS, Inc., or its subsidiaries in the United States or other countries. All other brand, product, service, and feature names or trademarks are the property of their respective owners.

Limit of Liability/Disclaimer of Warranty: While the publisher and author have used their best efforts in preparing this book, they make no representations or warranties with respect to the accuracy or completeness of the contents of this book and specifically disclaim any implied warranties of merchantability or fitness for a particular purpose. No warranty may be created or extended by sales representatives or written sales materials. The advice and strategies contained herein may not be suitable for your situation. You should consult with a professional where appropriate. Neither the publisher nor author shall be liable for any loss of profit or any other commercial damages, including but not limited to special, incidental, consequential, or other damages.

For general information on our other products and services or for technical support, please contact our Customer Care Department within the United States at (800) 762-2974, outside the United States at (317) 572-3993 or fax (317) 572-4002.

Wiley also publishes its books in a variety of electronic formats. Some content that appears in print may not be available in electronic formats. For more information about Wiley products, visit our web site at www.wiley.com.

Library of Congress Cataloging-in-Publication Data is available.

ISBN: 978-1-119-10344-8

Cover design: Wiley
Cover image: @Hramovnick/Getty Images

Printed in the United States of America.

V006893_042718

CONTENTS

PREFACE

ELECTRIC MACHINES are being used in wide and novel applications throughout the world, driven by the need for greater power efficiency in the transportation, aerospace and defense, and industrial automation markets. The automotive sector is driven by the need for hybrid and electric vehicle technology to meet ever-stringent miles-per-gallon standards. The aerospace and defense sectors are focused on replacing existing power transfer technologies in an aircraft such as the central hydraulic system, with fault-tolerant electric power, where major subsystems such as engine starting, primary flight control actuation, pumps, and braking would be controlled and driven electronically. In the US industrial sector, over 40 million electric motors convert electricity into useful work in manufacturing operations. Industry spends over $30 billion (US) annually on electricity, dedicated to electric motor-driven systems that drive pumps, fan and blower systems, air compression, and motion control. Globally, 42% of all electricity is used in power industries, where two-thirds of this is consumed by electric motors. There is a clear global demand for a comprehensive design methodology to support these new applications and satisfy power efficiency requirements.

With the present trend of global industrial automation, the application of electric drive systems (including power electronics and drive control) is expected to grow rapidly in the next decade. In the automotive sector, the utilization of power electronics and their control to drive electric motors can significantly contribute to control environmental pollution. In addition, intensive environmentally clean photovoltaic and wind energy resources also show a bright future.

As part of electric drive systems, the power semiconductor devices at the heart of modern power electronics are under continuous development. The improved technology in semiconductor processing, device fabrication, and packaging to produce high-density, high-performance, high-reliability, and high-yielding microelectronic chips, together with new semiconductor material discovery, made possible significant reduction in energy consumption, driving these systems to an incredible economical price.

Without doubt, these achievements force the control strategy techniques to evolve rapidly to the newly created drive conditions and adapt to the overall systems performance requirements. In recent years, soft switching converters became the center of interest when compared with more conventional hard switching converters due to their major advantages such as:

- Minimization of switching loss
- Improved efficiency

- Improved reliability due to soft stress
- Reduced electromagnetic emission

The continuous growing interest in the electric drive area relates to the intelligent power electronics modules, where the power and the control are embedded in the same package and interface directly with logic signals. For variable frequency drives, the converter modules and control are mounted directly on the machine for the low and medium power applications.

READERS' ADVANTAGE

The book is mainly addressed to design engineers, application engineers, technical professionals, and graduate engineering students with a strong interest in electric machines and drives.

The comprehensive design approach described in this book supports new applications required by technologies, sustaining high drive efficiency. The highlighted framework considers the electric machine at the heart of the entire electric drive. The book delivers the multiphysics know-how based on practical electric machine design methodologies. Simulation by design concept elevated in the book constitutes the new paradigm that frames the entire highlighted design methodology, which is described and illustrated by various advanced simulation technologies.

Which Design Problems Are We Trying to Solve?

Throughout this book, we apply knowledge of design best practices into multiphysics and multidomain simulation processes to address a complete electrical machine and drive design.

In the face of global competition, electric machine manufacturers, like manufacturers in most industries, are searching for ways to reduce cost, optimize designs, and deliver them quickly to market. Companies able to achieve these objectives hold a competitive advantage in the marketplace. The ability to predict design performance with simulation software without the time and expense of constructing prototypes plays a significant role in creating this competitive advantage.

Several computation approaches are available to predict electric machine performance, including classical closed-form analytical analysis, lumped parameter models based on the determination of detailed parameters from finite element analysis, and nonlinear time-domain finite element analysis. Each method has advantages and disadvantages. Selecting the best method may not be straightforward because it requires the user to understand the differences among the calculation methods. The fundamental issue differentiating these methods is the trade-off among model complexity, accuracy, and computing time. Engineers use a combination of these calculation techniques as the optimal solution to simulate electric machine performances.

What Motivated Us to Write This Book?

This collaborative work brought together a group of experts from both academia and software industry with strong expertise on electrical machine design and manufacturing. The main idea that fueled our initiative and commitment to make this project a reality was to bring back to the engineering and academic communities a comprehensive expertise and validated know-how on designing electrical machines by simulation.

Why Simulation by Design?

The advancements in modern digital computers brought CAD (computer-aided design) and CAE (computer-aided engineering) tools at the heart of virtual prototyping, reducing the time to design and market and saving cost by reducing and eliminating the physical prototyping need. The embedded 3D-physics design into drive system coupling with the power electronics and control algorithms enables the electric drive community to accurately predict the efficiency and performance of the electrical machine at the heart of the entire drive system.

Without doubt, the design of a simulation model—a virtual prototype—can help tremendously the engineers to build confidence on validating the required technical specifications making critical decisions on design realization and understanding the level of design complexity considering inter-dependencies and design parameter variations, and collaboratively to examine strategic choices for optimization and robustness.

CHAPTER DESCRIPTION

Chapter 1: Basics of Electrical Machines Design and Manufacturing Tolerances

This chapter discusses fundamental aspects of the state-of-the-art design process and includes examples from industrial practice and case studies to introduce basic concepts and methods. This chapter emphasizes the basic steps in designing a typical electrical machine using power traction application as an example. The chapter starts with magnetic sizing steps and it extends the basic design to thermal constraints. Typical electric motor characteristics used in traction applications such as efficiency map of standardized driving cycles are considered to highlight the electric motor sustainability on dynamic performance. The chapter concludes with the robust design analysis framing a methodology that applies stochastic analysis to study manufacturing tolerances.

Chapter 2: FEM-Based Analysis Techniques for Electrical Machine Design

In this chapter, a detailed description of finite element method (FEM) employed in ANSYS Maxwell software is presented. The numerical technique developed to account for eddy currents in conductive domains on configurations that involve rigid motion is presented, the numerical technique related to multiply connected regions is highlighted, and it also presents the algorithms used for nonlinear iterations and strategies to accelerate the nonlinear convergence. Filed-circuit coupling technology is explained and specific algorithms used to reduce the computation time to reach steady-state conditions are described. High-performance computing (HPC) is a key technology, increasing the capacity of solving large design spaces and reducing significantly the total time computation by solving the time steps on magnetic transient problem simultaneously rather than sequentially. All technologies highlighted in this chapter are explained through sets of case studies.

Chapter 3: Magnetic Material Modeling

This chapter introduces advanced magnetic material modeling capabilities employed in numerical computation. From isotropic nonlinear characteristics to anisotropic behavior corresponding to grain-oriented magnetic materials, the chapter describes the implementation aspects and detailed modeling techniques. Lamination topologies are considered based on special modeling technique with emphasis on core loss computation. Advanced magnetic modeling on vector magnetic hysteresis is presented and specific case studies are used to highlight the computational merits.

Chapter 4: Thermal Problems in Electrical Machines

In this chapter, the heat generation and extraction in electrical motors are investigated. Using the three thermal paths—conduction, convection, and radiation—an electromagnetic device can be cooled within the acceptable limits for the environment and corresponding application. A highly efficient electrical machine is required in most industrial fields, but the high efficiency is not telling us the full story of a good motor performance. The losses—electromagnetic and mechanical—must be dissipated from the machine into the ambient and the mode in which the cooling system manages to do that represents the key in a reliable and high-performance electrical machine. Within the chapter, the theoretical aspects of thermal management are illustrated with a state-of-the-art collection of practical examples for cooling electrical machines published in the literature.

Chapter 5: Automated Optimization for Electric Machines

This chapter discusses optimization as applied to electrical machine design. Some commonly used optimization methods are explained. Case studies illustrating the utility of systematic design optimization to compare different machine topologies, to develop design rules, and to quantify the effect of different design features are included.

Chapter 6: Power Electronics and Drive Systems

This chapter describes the entire drive system from semiconductor as the main component of any modern power electronics circuit to more complex topologies that include active components to rectify the energy, reduce harmonic distortions, and correct power factors in various drive systems. Electrical machines need drive systems to be correctly controlled if they need to be operated at variable speed. This can be achieved by modulating the energy flow to/from them. The chapter also highlights the need of multiphysics studies for such designs to account for thermal analysis under certain cooling conditions. For instance, inverter modules need a careful design approach as losses vary continuously during normal operation. Poor thermal management can lead to overheating and thus degrade the reliability of the components.

FRAMING THE MULTIPHYSICS DESIGN METHODOLOGY

The electric machine is a very complex device, being multidomain by nature involving electromagnetics, thermal, and mechanical aspects. The multiphysics methodology built around the core of electric machine design encompasses a systematic approach to develop a platform where comprehensive analysis is the key to understand and design a complex drive system to predict their performances and analyze their robustness. The multiphysics simulation technology enables users to design, analyze, and deliver efficient, optimized electric machine and drive designs.

As shown in Figure 1, the first step in the overall workflow is to develop design requirements. Those requirements may be created within a particular design organization, or they may be provided from a purchaser of the electric machine. Requirements may include machine speed, output power, input power, torque, efficiency, thermal properties, weight, size, etc. At this stage, motor sizing and model creation take place,

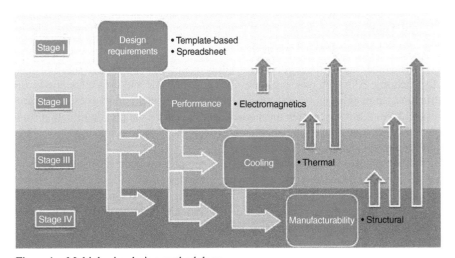

Figure 1   Multiphysics design methodology.

where many motor configurations may be considered. Often, engineers will use classical, closed-form analytical methods to select appropriate motor configurations that will meet requirements. In a similar manner, both magnetic and thermal designs can be evaluated using template-based approaches that are using such closed-form analytical methods. At the end of this stage, the designer acquires knowledge about the most suitable motor topology to fulfill the level of technical specifications with a degree of confidence on practical realization.

At the second stage of the workflow with the set of knowledge already acquired, accurate and detailed motor studies using 2D and 3D finite element analysis are performed. This important step in the design process further qualifies electromagnetically the topologies selected by the magnetothermal sizing analysis. Various design characteristics are numerically evaluated employing cutting-edge techniques, for example, permanent magnet demagnetization due to irreversible temperature effect, power electronics switching loss effect on electric motor core-loss, efficiency and power loss maps.

The thermal study can be developed similar to electromagnetic analysis as a separate design simulation or in connection with electromagnetic solution. With feedback from electromagnetic losses from either template-based solution or finite element analyses, the thermal study can be migrated from simple temperature rise computation based on thermal conduction to more complex studies involving computational fluid dynamics (CFD), where convection and radiation are considered. In such configurations, detailed cooling systems can be evaluated and optimized. Speeding up the entire thermal profile prediction is a key in the overall design process, that provides with design alternatives as,

- Creating an equivalent thermal model (ETM) for motor topology to be used within CFD environment to build around it the physical cooling system configuration with focus on outside enclosure optimization analysis or,

- Coupling detailed finite element-based integrated losses with thermal simulation model to accurately predict temperature profile, fluid flow, and its velocity within various parts of the entire assembly.

The final stage of the proposed workflow relates to mechanical design and manufacturability. Although this step follows electromagnetic and thermal analyses, the structural simulation can be performed any time during the multiphysics design process with emphasis on specific mechanical analyses, for example, deformation studies, noise vibration, and structural dynamics analysis, to more complex induced thermal stress and magnetostrictive analyses.

This simulation framework allows the engineer to understand the electrical, thermal, structural, and acoustical behavior of the design, considering electric motor as an independent component or part of an electric drive system including power electronics. Finally, the motor design is considered in the broader context of its power control unit and integration with other systems.

The flexibility of such a design flow is provided by the data exchange among all physics involved, providing various design adoption alternatives.

The multiphysics design flow can be further detailed at each and every individual stage. In spite of this granularity, the Chapter 1 will focus on Stage I regarding the generic design flow for topology selection during a motor design to examine the process of basic design.

Marius Rosu
Ping Zhou
Dingsheng Lin
Dan Ionel
Mircea Popescu
Frede Blaabjerg
Vandana Rallabandi
David Staton

ACKNOWLEDGMENTS

THE AUTHORS are grateful for the tireless efforts, assistance, and guidance of the Wiley-IEEE Press editorial staff who brought to reality this book project.

We are thankful to the many colleagues who provided technical insights, comments, and suggestions.

We are especially indebted to our partners, collaborators, and customers who supported and diligently demanded continuous progress on software technology enabling us to thriving innovation.

We owe a great debt of gratitude to our families for their unconditional support and continuous encouragements.

This book could not be written to its fullest without ANSYS Inc.'s support and continuous engagement for which we are grateful.

CHAPTER 1

BASICS OF ELECTRICAL MACHINES DESIGN AND MANUFACTURING TOLERANCES

1.1 INTRODUCTION

Recent progress in the area of electric machines, including new materials, manufacturing technologies, and conceptual topologies, require a systematic design approach to ensure improved performance and/or reduced cost for new developments. This chapter discusses fundamental aspects of the state-of-the-art design process and includes examples from industrial practice and case studies in order to introduce basic concepts and methods.

Traditionally, the core of electric machines is manufactured by punching and stacking thin laminations of cold rolled or silicon steel, as illustrated in Figure 1.1 [1]. Even with the rotor laminations nested inside the stator ones, the process results in a relatively large amount of steel being scraped due to the slots and outer stator profile. Depending on the machine type and design topology, in the rotor, permanent magnets (PM) maybe inserted or attached to the core, an electrically conductive cage maybe die-cast from aluminum, for example, or a winding (not applicable for the design shown in Figure 1.1) maybe inserted. The stator typically incorporates a collection of coils made of conductive wires separated by electrical insulation and forming a winding. A multiphase distributed winding, such as the one exemplified in Figure 1.1, maybe manufactured by automatically producing the coils and then inserting them into the core, in a process that has high through output, and results in a relatively high ratio of net conductor per slot area, that is, slot fill factor, but also yields relatively large end coils.

In order to reduce the scrap of laminated steel, different punching arrangements may be employed. The example shown in Figure 1.2 is particularly advantageous for stator designs with relatively large tooth width and small tooth tips. Modules of core and concentrated coils, each wound around a tooth, can be organized with a single or multiple teeth, formed to shape, and then assembled to produce a stator, such as

Multiphysics Simulation by Design for Electrical Machines, Power Electronics, and Drives, First Edition.
Marius Rosu, Ping Zhou, Dingsheng Lin, Dan Ionel, Mircea Popescu, Frede Blaabjerg, Vandana Rallabandi, and David Staton.
© 2018 by The Institute of Electrical and Electronics Engineers, Inc. Published 2018 by John Wiley & Sons, Inc.

Stator core and winding

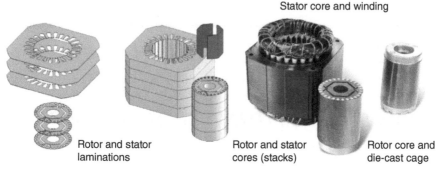

| Rotor and stator laminations | Rotor and stator cores (stacks) | Rotor core and die-cast cage |

Figure 1.1   Typical steps for the manufacturing of an electric machine, in this case a line-fed permanent magnet (PM) synchronous motor, which includes in the rotor a die-cast aluminum cage.

the three-phase 18-slot design example from Figure 1.2, which is suitable to be used together with a 16-pole PM rotor [2, 3].

At a first look, concentrated windings, especially in a segmented-modular configuration, tightly packed with a high slot fill factor and short end coils maybe superior to more conventional distributed winding machines, particularly for low-speed applications. Nevertheless, before drawing such generic conclusions, systematic comparisons taking into account the power and speed rating, losses, including winding and core components, the electronic controller should be performed following, for example, a large scale automated design process as described in another chapter of the book.

A stator core and winding, and a rotor incorporating a shaft, are assembled together with other components, including bearings, end caps, and frame and terminal box to produce an electric machine, as exemplified in Figure 1.3 for a general-purpose three-phase squirrel-cage induction motor and in Figure 1.4 for PM

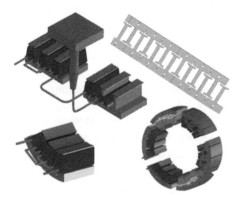

Figure 1.2   Example of stator modular construction with segmented core and concentrated coils forming a multiphase winding [2, 3]. Versions with a single tooth and coil modules are also possible.

Figure 1.3   Exploded view of a general-purpose National Electrical Manufacturers Association (NEMA) frame squirrel-cage induction motor. *Courtesy of Regal Beloit Corp.*

synchronous motors with stator concentrated and distributed windings, respectively [4]. It should be noted that the use of PM technology typically results in a higher power density than available from induction machines.

A most successful example of combining advanced design techniques, high performance magnetic materials, and enhanced cooling is represented by the recent development of a 100 hp motor for Formula E racing cars [5,6]. This machine shown in Figure 1.5, which employs an 18-slot 16-pole spoke IPM configuration, sets a record for electric traction motors of comparable rating, achieving almost twice the specific torque density per unit of active mass (Nm/kg) than the motor powering the latest generation of the Nissan Leaf electric vehicle.

1.2   GENERIC DESIGN FLOW

In the overall economy of the electric machine design process, virtual prototyping is the most important collection of designing stages when considering design validation

Figure 1.4   PM synchronous motors with interior permanent magnet (IPM) rotors and concentrated (left) and distributed windings, respectively [4]. *Courtesy of Regal Beloit Corp.*

Figure 1.5   Record-breaking ultra-high density torque 100 hp spoke-type IPM motor for Formula E racing cars showing, from the left, magnetic field in the motor cross section, axial cut view and photo [4,5]. *Courtesy of Equipmake, Ltd.*

before committing to making a physical prototype. The obvious procedures utilized throughout this process combine efficient numerical techniques providing with highly accurate, comprehensive multiphysics simulation designs including power electronics and control software to predict the electric drive performance at the heart of entire system.

The electrical machines are part of the electromechanical energy converters group. This means that any electrical machine, be it rotating, or linear movement, with AC or DC supply current, can operate as:

- Motor: conversion of the input electrical energy into output mechanical energy
- Generator: conversion of the input mechanical energy into output electrical energy
- Brake: absorption of both mechanical and electrical energy

In electromechanical terms, the main inputs and outputs of the electrical machines are the supply voltage and current versus the torque (or force) at the shaft and rotational (or linear) velocity.

The main design flow is shown in Figure 1.6. The initial design is addressed at the level of sizing stage when physical dimensions are established based on magnetic and thermal feedback. This is a specialized stage that allows designers to quickly create a geometric model of the machine, calculate its performance, and make sizing decisions. Once the initial design is completed, detailed numerical techniques (2D and 3D) are necessary to increase the accuracy of such designs by simulation. Such simulations are referred as physics-based designs whether the numerical technique relates to the electromagnetics simulation by finite elements or fluid dynamics solution by finite volumes or structural simulation by finite element analyses.

In this chapter, we will focus on the basic steps in designing a typical electrical machine for power traction application.

1.3   BASIC DESIGN AND HOW TO START

We will consider the specifications shown in Table 1.1 for a power traction application. If someone will ask what parameter would be the most important to determine

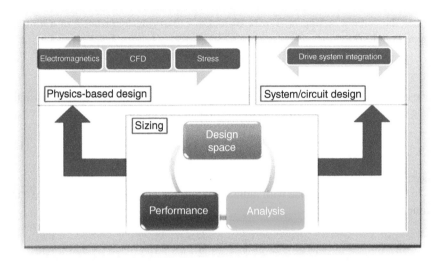

Figure 1.6   Multiphysics and multidomain electric machine design flow.

the physical dimensions of an electrical machine, we can definitely answer that the electromagnetic toque capability, which can be extracted from the design specification related to a certain application, will generate enough information to produce an initial magnetic design. And yet this is probably at the same time the most exquisite and intrigued statement one could deliver on such a complex design demand [7, 8].

Simply, following this rationale, the output torque is proportional to rotor volume and tangential force acting on rotor surface known as shear stress. The shear stress is proportional to the product of electric and magnetic loading. Besides electromagnetic torque, the rotor aspect ratio, stator slot diameter ratio, number of poles, and rotational speed are important parameters for design decisions.

TABLE 1.1   Power traction application specifications

Parameter	Unit	Value
DC supply voltage	V	400
Maximum DC current	A	700
Maximum line AC current	Arms	900
Peak output power	kW	235
Peak torque	Nm	330
Base speed	rpm	7500
Maximum speed	rpm	15,000
Continuous power at base speed	kW	80
Continuous torque at base speed	Nm	100
Cooling system	N/A	Liquid
System envelope volume	mm	$270 \times 270 \times 270$

The fundamental relationship for torque prediction is the equation which defines the force F acting on a wire of length L carrying a current I in a uniform magnetic field B as

$$F = B \cdot I \cdot L \tag{1.1}$$

If we consider a uniformly distributed conductor per meter, then we can introduce the shear stress as the ratio between force and area of conductive region. Thus,

$$\sigma = \frac{F}{\text{Area}} = B \cdot A \tag{1.2}$$

where B is also known as magnetic loading and A is known as electric loading.

In any electric machine with radial field concept with the rotor diameter D, the torque is introduced by

$$T = F \cdot \frac{D}{2} \tag{1.3}$$

In more general terms, the main parameter defining an electrical machine is the torque or force depending on the device that has a rotational or linear movement, respectively. In this design example, we are dealing with a rotational movement, for which the expression that provides the sizing **electromagnetic torque** is given by

$$T = \frac{\pi^2}{4\sqrt{2}} \cdot k_{w_1} A B D^2 L_{stk} \tag{1.4}$$

where k_{w_1} is the fundamental winding factor, L_{stk} is the axial active length.

A is the **electric loading**: number of ampere-conductors per meter around the stator surface that faces the airgap.

$$A = \frac{\text{Total ampere–Conductors}}{\text{Airgap circumference}} = \frac{2mN_{ph}I_{RMS}}{\pi D} \tag{1.5}$$

where N_{ph} represents the total number of turns/phase, I_{RMS} is the RMS phase current and m is the system number of phases.

B is the **magnetic loading:** average flux-density over the rotor surface. If the flux-density is distributed sinusoidal, the fundamental magnetic flux/pole (there are $2p$ magnetic poles in the electrical machine) is

$$\phi = B \cdot \frac{\pi D L_{stk}}{2p} \tag{1.6}$$

We notice from (1.4)–(1.6) that the torque, electric, and magnetic loading depend on the machine volume in the airgap.

Considering also the lossless **energy conversion** law for an m-phase AC machine,

$$\text{Electric power} = mEI = \text{Mechanical power} = T \cdot \frac{\omega}{p} \tag{1.7}$$

where E is the **induced voltage** (back electromotive force—emf), ω angular velocity. The induced voltage can be further developed as

$$E = \frac{\pi^2}{\sqrt{2}} \cdot \frac{fk_{w_1}N_{ph}BDL_{stk}}{p}, \tag{1.8}$$

where f is the fundamental frequency.

It is also worth noting the relation between fundamental frequency, f, number of pole pairs, p, and rotational speed n, where rotational speed is represented in revolution-per-minute (rpm),

$$f = \frac{n \cdot p}{60} \tag{1.9}$$

In sizing an electrical machine, we can use a combined parameter, that is, the torque ratio per volume based on relation (1.4) and knowing the rotor volume, $V_{rotor} = \frac{\pi D^2 L_{stk}}{4}$, then

$$TRV = \frac{T}{V_{rotor}} = \frac{\pi}{\sqrt{2}} \cdot k_{w_1} A \cdot B \tag{1.10}$$

The above equation reflects the electromechanical nature of the energy conversion, as A is proportional to the current in the machine windings, while B is proportional to the magnetic flux. Hence, the electric energy that is proportional to the product of magnetic flux ϕ and current I is converted to mechanical energy and relates to the torque production.

Table 1.2 summarizes the typical values for torque ratio per volume in various electrical machine types.

Another simple initial consideration is related to the current density in the electrical machine, depending on the cooling system (as represented in Table 1.3).

For the considered specifications, (as shown in Table 1.1) and assuming a split ratio (i.e., rotor/stator outer diameters) of 0.55, and that the active length of the machine occupies maximum 80% of the overall maximum axial length, we can estimate peak TRV as:

$$TRV_{peak} = \frac{330}{\dfrac{\pi \cdot (0.55 \cdot 0.27)^2 \cdot 0.27 \cdot 0.8}{4}} \cdot 0.001 = 88.20 \text{ kN/m}^3$$

TABLE 1.2   Typical values for TRV continuous operation in electrical machines

Electrical machine type	TRV (kNm/m³)
Totally enclosed motors—low energy magnets (ferrite)	5–15
Totally enclosed motors—sintered rare-earth magnets (NdFeB, SmCo)	15–40
Totally enclosed motors—bonded rare-earth magnets (NdFeB, SmCo)	10–20
Medium power (>5 kW) industrial induction motors	5–30
Servomotors	15–50
Aerospace machines	30–75
Liquid cooled machines	75–250

TABLE 1.3 Typical current density values in electrical machines — continuous operation

Cooling system	Current density (A/mm²)	Current density (A/in²)
Totally enclosed non-ventilated	1–5	650–3250
Open air, fan cooled	5–10	3250–6500
Forced liquid cooled	10–30	6500–20,000

Similarly, continuous TRV as:

$$TRV_{cont} = \frac{100}{\dfrac{\pi \cdot (0.55 \cdot 0.27)^2 \cdot 0.27 \cdot 0.8}{4}} \cdot 0.001 = 26.73 \text{ kN/m}^3$$

The above values suggest that we will select a sintered rare-earth permanent magnet machine, that will need forced liquid cooled system, while the overall dimensions of the machine can use a rotor diameter, $D \sim 150$ mm, that is, 0.55×270 mm.

For axial length, we will select the value, $L_{stk} = 150$ mm, hence the machine rotor axial view will have a square profile ($D \sim L_{stk}$).

Assuming the ratio of stator and rotor dimensions is kept constant, increasing the rotor diameter will increase the stator slot depth, the electric loading, and the torque ratio per volume. Important advantages include smaller stator stack length, larger shaft diameter which implies large inertia and higher critical speed.

Similarly, under the same assumptions, the reduction on rotor diameter decreases the rotor inertia enhancing the dynamic response. Such configuration generates lower rotor mechanical stress at high speeds and reduces the end-turn copper losses. Lower diameter of the shaft implies smaller diameter for bearings which sustain higher speed operations at lower friction losses.

1.3.1 Stator and Rotor Topologies

When designing a brushless permanent magnet machine, there are several configurations that can be selected for stator and rotor assembly. At this stage, we do know the overall diameters and axial length only.

Rotor Magnetic Pole Pairs (p)

First step in further design procedure is to select the number of magnetic pole pairs. Based on the imposed maximum rotational speed, 15,000 rpm, and equation (1.9), it is possible to shortlist few suitable numbers. As derived from Table 1.4, we consider the maximum fundamental frequency to be 1000 Hz and this limits the suitable number of pole pairs to 1, 2, 3, and 4.

Considering the suitable number of pole pairs and the higher number of pole pair advantages:

- a higher torque ratio per volume due to the reduced leakage magnetic flux
- lower dimension of the stator back iron

TABLE 1.4   Variation of the fundamental frequency with number of pole pairs

Rotational speed (rpm)	Pole pairs	Poles	Fundamental frequency (Hz)
15,000	1	2	250
	2	4	500
	3	6	750
	4	8	1000
	5	10	1250
	6	12	1500

- reduced end-winding dimensions with implications on reduced copper losses and material consumption

However, higher number of poles generates higher permanent magnet resistive losses due to eddy-current effects induced in these conductive materials especially when magnets are mounted on the surface of the rotor.

Higher number of poles requires a short pole pitch which is well suited to use in concentrated windings. For distributed windings, the utilization of short pole pitch requires larger number of slots which can increase the mechanical vibrations, hence the acoustic noises.

The reluctance torque is proportional to the inductance of the winding which is inversely proportional to the square of the number of poles. Therefore, at higher number of poles the reluctance torque is significantly reduced.

Considering the advantages versus the disadvantages of higher specific core losses and converter losses due to higher frequency, still leads to the conclusion that the number of pole pairs equal to 4 is a good design choice.

Hence, we decide to have a rotor with four pole pairs (eight magnetic poles).

Rotor Poles Topology

The eight rotor poles can be designed with magnets mounted on the surface of the rotor (SPM—surface permanent magnet motor), or embedded in the rotor magnetic core (IPM—interior permanent magnet motor). For a traction application, the IPM topology has certain advantages as the magnet blocks can be of rectangular prism shape, better protected against demagnetization due to faults and high temperature profiles, and allows a more efficient field weakening operation.

The V-shape topology shown in Figure 1.7 is chosen.

Stator Number of Slots and Winding Topology

A balanced stator lamination and winding requires that certain rules are respected when deciding the number of slots:

- Slots/pole pairs >1 (no emf is possible otherwise)
- Slots/phases/parallel paths = Integer

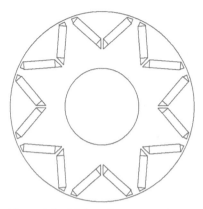

Figure 1.7   Eight-pole rotor topology.

- Pole pairs/parallel paths = $2k$, where k = Integer
- Slots/phases/greatest common divisor (GCD; Slots, pole pairs) = Integer

Following the above rules, a three-phase, 8-pole rotor can be paired with a stator with 12, 24, 36, 48, 72, 84, 96 slots.

In relation to the stator bore that is in the region of 150 mm, from the manufacturing point of view and cooling efficiency, **the number of stator slots is set to be 48**.

Figure 1.8 illustrates the proposed radial view of this electric motor design.

The winding is selected (as seen in Figure 1.9) to be:

- Three-phase, star connection
- Single coil layer (easier to implement and with high copper slot fill factor)

Figure 1.8   48 slots/8-pole IPM machine configuration.

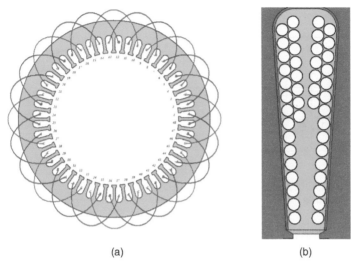

(a)   (b)

Figure 1.9   (a) Winding pattern; (b) winding slot detail.

- Short pitch coil, that is, 1–6, compared to full pitch coil 1–7; this will eliminate certain higher-order harmonics
- Parallel paths (branches) are 2
- Turns/coil = 3
- Strands in hand = 7
- Wire gage = 1.3 mm

1.3.2   Drive Topologies

As schematically shown in Figure 1.10, either insulated gate bipolar transistors (IGBTs) or metal–oxide–semiconductor field-effect transistors (MOSFETs) are typical semiconductor devices used as switches on three-phase H-Bridge configurations.

The maximum peak line–line voltage equals the DC bus battery, and hence the maximum RMS line—line voltage $V_{RMS} = V_{DC}/\sqrt{2}$, where V_{DC} is the DC bus supply voltage. In reality, this is typically lower and dependant on modulation strategy.

The three-phase inverter controls the current by applying appropriate voltage at the machine terminals.

The terminal voltage must overcome the emf and voltage drop across the inductance to inject the correct voltage.

Field weakening is used to extend the speed range of the machine when the emf is greater than the maximum available terminal voltage.

Table 1.5 gives the ratio between inverter output RMS voltage and the input DC bus voltage for various pulse width modulation (PWM) control strategies.

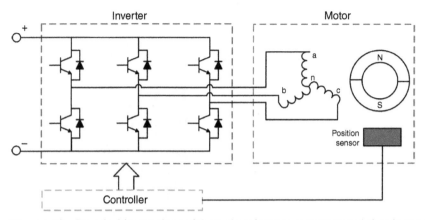

Figure 1.10 Electric drive topology of three-phase inverter, star-connected electric motor with controller.

1.3.3 Torque Speed Characteristics

In a wide speed range operation, the torque speed characteristics for IPM machines consist of three modes (Figure 1.11):

- In Mode I, the torque is limited by the current.
- In Mode II, the maximum current is used and the current angle is advanced to suppress the voltage.
- In Mode III, the current value converges toward the current value for theoretical infinite speed of λ_m/L_d, where λ_m is the rotor magnetic field and L_d is the D-axis inductance of the machine.

1.3.4 Electromagnetic Simulation Using Finite Element Analysis

The simulation presented has been achieved using ANSYS Maxwell which is capable of performing rigorous performance calculations of the machine including

TABLE 1.5 PWM strategy

PWM strategy	Line–line (RMS)/DC bus voltage ratio	Line–line (peak)/DC bus voltage ratio
Six-step 180	0.780	1.10
Hexagon tracking: piecewise linear	0.7446	1.05
Hexagon tracking: secant	0.7418	1.05
Third harmonic injection	0.707	1.00
Six-step 120	0.675	0.95
Maximum linear range of sine/triangle	0.612	0.87

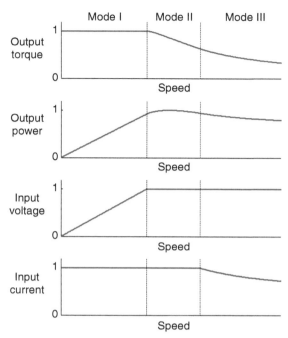

Figure 1.11   Operation modes in IPM machines.

the motion-induced physics caused by linear translational and rotational motion, advanced hysteresis analysis, demagnetization of permanent magnets and other critical electromagnetic machine parameters.

Starting exclusively on electromagnetic design flow narrowing the design space with ANSYS RMxprt, the magnetic transient design setup is automatically generated as either 2D or 3D designs. The benefit of 2D symmetry is due to radial magnetic field concept topology for most of electrical machines used in different applications. However axial field or transversal field topologies require 3D designs. Nevertheless, for accuracy reasons, 3D designs are required even if radial field topologies are considered when end effects, multiaxial segmented permanent magnets, or skewing topologies are employed. Although ANSYS Maxwell package is a general finite element analyses tool, its capabilities enable designers to customize and apply very specific analyses for electrical machines as D-Q solution computation (Figure 1.12), or employ even more complex algorithms as maximum torque per ampere unit (MTPA) control strategy (Figure 1.13) used to construct maps of efficiency and losses.

On the motor topology selected and sized based on the concepts developed in Section 1.3, we employed magnetic time-stepping (transient) analysis to predict more accurately the dynamic performance of such IPM design. Figure 1.14 shows field distributions on the surface of rotor and permanent magnets while Figure 1.15 and Figure 1.16 represent the transient data of running torque and induced voltage (back emf) profiles, respectively.

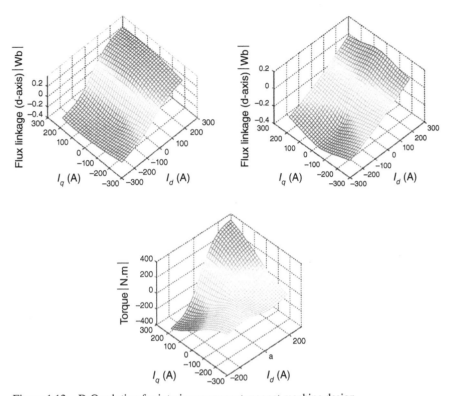

Figure 1.12   D-Q solution for interior permanent magnet machine design.

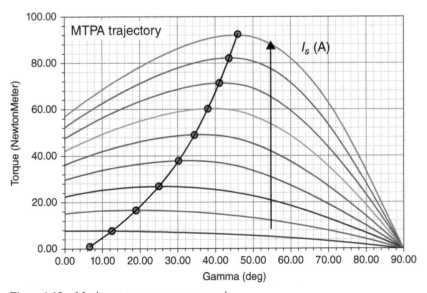

Figure 1.13   Maximum torque per ampere unit.

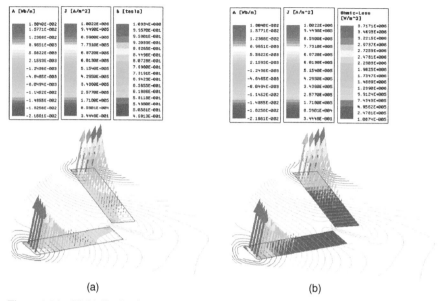

(a)   (b)

Figure 1.14   Field distribution on rotor and interior permanent magnets at the time instant of maximum input phase current:
(a) flux lines with magnetic field and induced current density distribution on PM,
(b) flux lines with power loss and induced current density distribution on PM.

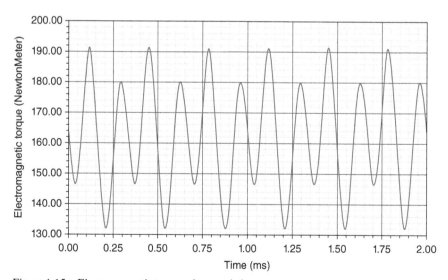

Figure 1.15   Electromagnetic torque characteristic.

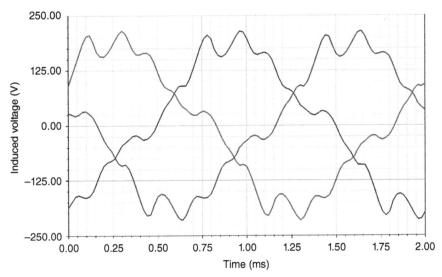

Figure 1.16 Induced voltage profile.

1.4 EFFICIENCY MAP

The AC brushless motors can be controlled using various control strategies. The most common strategies are:

- Maximum torque per ampere
- Maximum efficiency
- Unity power factor
- Constant phase advance angle, for example, $I_d = 0$, while I_q = variable amplitude
- Constant power loss

The first two methods are described in Figure 1.17:

Maximum torque per amp		Maximum efficiency	
Minimize	$I_s = \sqrt{I_d^2 + I_q^2}$,	Minimize	$W_{total} = W_{cu} + W_{fe} + W_{magnet}$,
Subject to	$T_{shaft} - T_{demand} = 0$	Subject to	$T_{shaft} - T_{demand} = 0$
and	$V_{lim} \geq 2\pi f \sqrt{\psi_d^2 + \psi_q^2}$.	and	$V_{lim} \geq 2\pi f \sqrt{\psi_d^2 + \psi_q^2}$.

Figure 1.17 Control strategies for AC brushless PM machines.

Using MTPA control strategy and the power converter limits ($V_{DC} = 400$ V, $I_{DC} = 500$ A), it is possible to calculate torque–speed curves for the whole speed range. For each operation point, the following algorithms are considered:

• flux-linkages and inductances are calculated using 2D nonlinear transient finite element analysis model (e.g., ANSYS Maxwell)

• magnet losses are calculated considering the induced eddy-currents in the magnet blocks

• core losses are calculated via post-processing in finite element analysis with the flux-density distribution and applying statistical curves for the specific losses (e.g., Bertotti or Steinmetz models)

• Joule losses are calculated analytically for the DC component, and numerically for AC component based on finite element analysis

• mechanical/friction losses are determined using previous experience or measured data

With the amount of data created for the whole torque speed range, it is useful to display the loci of the efficiency isolines. These graphs are defined as efficiency maps and used to interpret the machine capability for entire operation spectrum, so that the designer can identify the speed and torque range where the efficiency is maximized, for example.

If the drive cycle is known, the operation points of this cycle can be superimposed on the efficiency map and thus observe if the machine was designed for optimal performance.

In automotive applications, there are standard drive cycles. Worldwide, the drive cycles presented in the Table 1.6 are representative:

TABLE 1.6   Standard drive cycles

Average speed	Distance	Time (s)	Description	Region	
19.59 mph	7.45 m	1369	Typical city driving conditions	USA	UDDS
48.3 mph	10.26 m	765	Typical highway conditions under 60 mph	USA	HWY
48.4 mph	8.01 m	569	High acceleration aggressive driving schedule	USA	US06
62.6 km/h	6.96 km	400	Extra-urban cycle	Europe	EUDC
59.5 km/h	6.6 km	400	Extra-urban (low powered vehicle)	Europe	EUDCL
18.4 km/h	0.99 km	195	Elementary urban cycle	Europe	ECE
17.7 km/h	0.66 km	135	10 mode cycle	Japan	JPN10
33.9 km/h	2.17 km	231	15 mode cycle	Japan	JPN15
25.6 km/h	6.34 km	892	Combined 10/15 mode cycle	Japan	JPN10.15

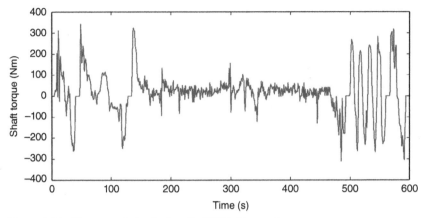

Figure 1.18   Torque variation in time for US06 drive cycle.

For the specific design in this chapter, we can consider the high acceleration aggressive drive cycle, US06 (USA). The detailed torque and power variation with rotational speed for this cycle are given in Figures 1.18 and 1.19.

Now, considering the ambient operation at 70°C and using the MTPA control strategy with flux-linkages and currents calculated with the electromagnetic model previously developed, the following efficiency maps can be plotted in both motoring and generating modes with drive cycle points superimposed. The efficiency map is generated in this example using the LAB module from Motor-CAD™ software. The losses are extracted from ANSYS Maxwell 2D module.

Figure 1.20 shows a maximum efficiency of 96% in the region of 2000 rpm to 9000 rpm and up to 180 Nm. We notice that more than 60% of the driving points will be in the region with high efficiency.

For the base point, 100 Nm at 7500 rpm we get at the same fixed temperature of 70° as illustrated in Figure 1.21.

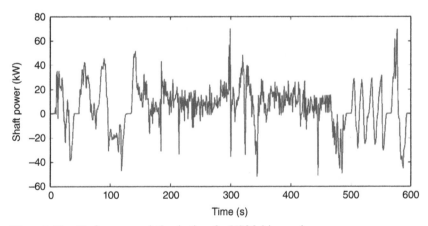

Figure 1.19   Shaft power variation in time for US06 drive cycle.

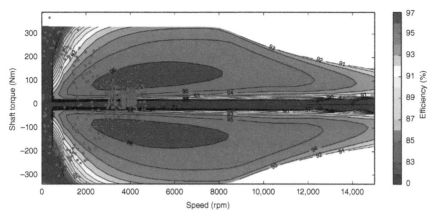

Figure 1.20   Efficiency maps in motoring and generating modes with superimposed US06 drive cycle points (blue dots).

Input Operating Point					Input:		
Torque Required	100	Nm	@ Speed	7500	RPM	⦿ Torque	○ Maximum Current

Speed (rpm)	7500	Total Loss (W)	3270.95	Reluctance Torque (Nm)	15.6773
Shaft Torque (Nm)	99.9999	Copper Loss (W)	959.962	Terminal Power (kW)	81.8107
Shaft Power (kW)	78.5397	Iron Loss (W)	2127.84	Voltage (Vphase[rms])	123.533
Efficiency (%)	96.0018	Magnet Loss (W)	22.427	Power Factor	0.9171
Id (Arms)	-96.3233	Mechanical Loss (W)	160.714	Flux Linkage D (Vs)	0.03732
Iq (Arms)	220.598	Electromagnetic Power (kW)	80.8507	Flux Linkage Q (Vs)	0.04049
Phase Advance (Elec deg)	23.5883	Electromagnetic Torque (Nm)	102.942	Vd (Vphase[pk])	-127.945
Is (Arms)	240.711	Magnet Torque (Nm)	87.265	Vq (Vphase[pk])	118.958

Figure 1.21   Estimated electrical machine parameters for rated operation point in Motor-CAD / LAB module (100 Nm at 7500 rpm).

The results highlighted in Figure 1.21, show that about two-thirds of the losses are in the iron region, more specifically in the stator iron.

1.5   THERMAL CONSTRAINTS

In Section 1.4, we have seen the electrical machine capabilities for whole speed and current range, considering a fixed temperature of 70°C. This value would correspond to an ambient temperature in the engine compartment of an electrical vehicle.

When an electrical machine is designed for the required performance, the thermal response analysis is essential.

The main aspects are:

- The power output of an electrical machine is strongly affected by its thermal performance because machine operating temperature limits the electric loading, $q = f(I)$
- Machine life, with "10°C half-life rule", that is, every 10°C increase in operating temperature cuts insulation life by half
- Permanent magnet materials: the performance of magnets decreases with temperature as the magnets will be demagnetized if they are overheated
- Copper loss is temperature dependent
- Mechanical, elevated temperatures can induce mechanical stresses due to thermal expansion
- Bearing failure
- Thermal fatigue during variable speed operation

The losses in any electromagnetic device will generate heat. A good performance of an electrical machine relies on an efficient heat extraction to the ambient environment via one or more of the three main mechanisms: conduction, convection, and radiation (for more details, see Chpater 5, the dedicated chapter to thermal analysis)

The designed electrical machine example has a high torque per volume ratio (TRV) and most of the losses are located on the stator assembly (winding and laminated core). Hence, a good cooling system must use liquid forced convection. As illustrated in Figure 1.22, we can design a housing water jacket with spiral channels.

Figure 1.22   Three-dimensional view of the stator assembly and cooling jacket.

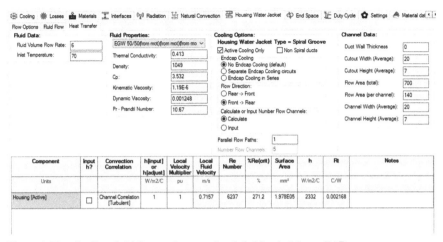

Figure 1.23   Cooling fluid flow and properties definition in Motor-CAD.

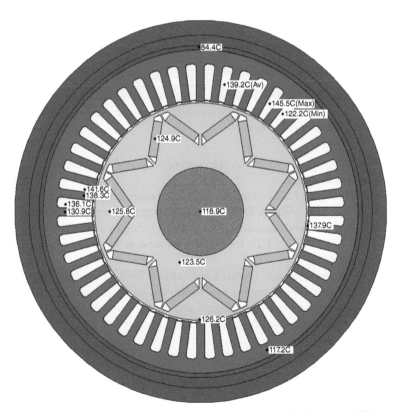

Figure 1.24   Radial view of the steady-state temperature distribution at 100 Nm at 7500 rpm operation point—based on thermal network.

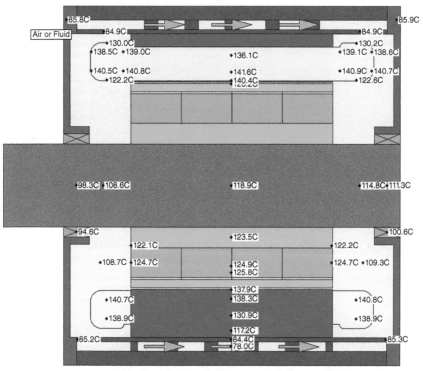

Figure 1.25   Axial view of the steady-state temperature distribution at 100 Nm at 7500 rpm operation point—based on thermal network.

Using Motor-CAD^{TM}, we can decide on the cooling channels' dimensions, on the cooling fluid, and the flow rate.

As observed in Figure 1.23, this machine is cooled using a housing spiral jacket with water ethylene–glycol mixture, 50–50%, fluid flow rate 6 L/min and inlet temperature of 70°C. The Reynolds number is 6237, which indicates a turbulent flow.

From Figure 1.24 to Figure 1.27, the thermal response at this load point will be using a lumped thermal network.

The computed characteristics reported in Figures 1.28–1.32 consider thermal envelope for continuous motor operation with maximum winding average temperature of 180°C and maximum permanent magnet average temperature of 140°C.

Figures 1.28–1.29 show that the proposed design is limited thermally by the magnet temperature.

1.6   ROBUST DESIGN AND MANUFACTURING TOLERANCES

Two important concepts for robust design and reliability analysis need to be introduced to better understand technical approaches will be further employed. To avoid

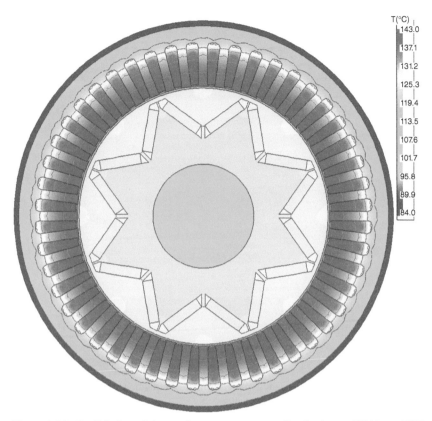

Figure 1.26   Radial view of the steady-state temperature distribution at 100 Nm at 7500 rpm operation point—finite-element thermal analysis.

further confusion, one can say that reliability analysis assesses the probability that various unpredictable design conditions yield unsustainable consequences, whereas robust design is the process based on which the required product specifications will be always satisfied in spite of variability and uncertainty.

Because a good design point is often the result of a trade-off between various objectives, the exploration of a given design cannot be performed by using optimization algorithms that lead to a single design point. It is important to gather enough information about the current design so as to be able to answer the so-called "what-if" questions, quantifying the influence of design variables on the performance of the product in an exhaustive manner. By doing so, the right decisions can be made based on accurate information, even in the event of an unexpected change in the design constraints.

For the sake of robust design representation on previously selected electric motor topology, Section 1.6.1 describes various simulation studies employing Design of Experiments (DOE), Response Surface, Parameter Correlation, Sensitivity and Six Sigma analyses.

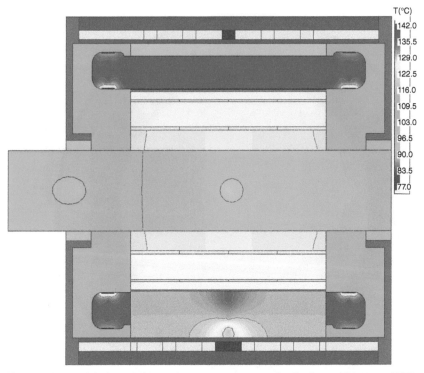

Figure 1.27   Axial view of the steady-state temperature distribution at 100 Nm at 7500 rpm operation point—finite-element thermal analysis.

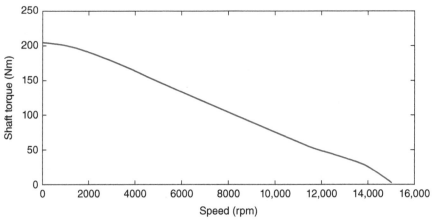

Figure 1.28   Continuous torque at average winding temperature <180°C and magnet temperature <140°C.

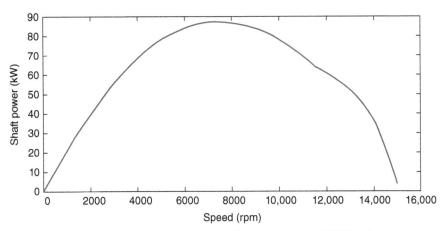

Figure 1.29 Continuous power at average winding temperature <180°C and magnet temperature <140°C.

1.6.1 Case Study

The proposed case study highlights meaningful steps on achieving robust design, enabling the engineers to account for uncertainties in product design and to determine how best to improve product reliability.

ANSYS Maxwell is used as finite element simulation tool to accurately predict the performance of the electric motor design while ANSYS Design Exploration provides with the robust design flow calling for an electromagnetic simulation on each design point defined by DOE. To ease the data transfer for enhancing the robust design with high level of flexibility, the flow highlighted in Figure 1.33 illustrates automatic connectivity among various components of the entire design analysis.

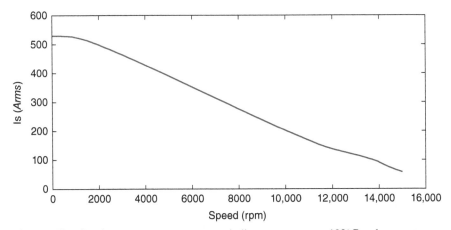

Figure 1.30 Continuous current at average winding temperature <180°C and magnet temperature <140°C.

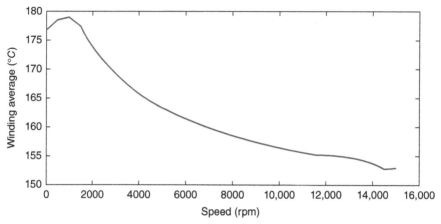

Figure 1.31   Average winding temperature variation with speed at steady-state thermal conditions.

To highlight the benefit of employing robust design analysis on electric machine design, the manufacturing tolerances are considered on the IPM motor topology highlighted in Tables 1.7 and 1.8.

Identify Key Design Parameters

The stator slot design has been selected (as seen in Figure 1.34) with parallel tooth configuration. Thus, only the top slot width (Bs2) varies while the bottom slot width (Bs1) is kept at a constant ratio of Bs2/Bs1 = 1.6. The rotor outer diameter as shown in Figure 1.34 will control the air-gap length, since the inner stator diameter (150 mm) is structurally constrained. The outer stator diameter is also fixed (220 mm) due to

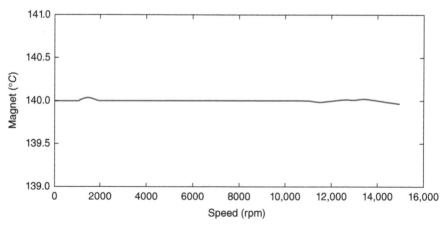

Figure 1.32   Magnet temperature variation with speed at steady-state thermal conditions.

housing and water jacket cooling configurations. Consequently, the stator slot height is fixed (22 mm) to maintain sufficient stator back iron thickness. A whole-coiled winding topology has been selected, as seen in Figure 1.35, with a fixed coil pitch (5) and 7 strands per conductor in two parallel branches distribution. Due to the minimum power density required to achieve the desired performance at base speed and maximum speed, both the length (50 mm) and width (6 mm) of IPM are fixed.

First part of such study describes the relationship between the design variables and the performance of the electric motor (Table 1.9) by using DOE, combined with Response Surface (RS). The main purpose of combined DOE with RS is to identify the relationship between the selected performance of the electric motor (P6 as average of electromagnetic torque and P7 as electromagnetic torque ripple) and the design variables (stator slot opening, stator slot width, air-gap length and the thickness of the IPM bridge). Practically, the design techniques are required to find the tolerance margins of the design variables in order to satisfy the optimized performance of the electric motor with certain confidence. Once the variation of the performance with respect to the design variables is known, it becomes easy to understand and identify all changes required to meet the requirements for the electric motor. For instance, once the response surfaces are created, the information can be shared in easily understandable terms: curves, surfaces, sensitivities. They can be used at any time during the development process without requiring additional simulations to test a new motor design instance.

Design of Experiments and Response Surface Analysis

In a process of engineering design, it is very important to understand what and how many input variables are contributing factors to the output parameters of interest. It is a lengthy process before a conclusion can be made as to which input variables play a role in influencing the output parameters. Designed experiments help revolutionize the lengthy process of costly and time-consuming trial-and-error search to a powerful and cost-effective statistical method.

A very simple designed experiment is screening design. In this design, a permutation of lower and upper limits of each input variable is considered to study their effect to the output parameters of interest. While this design is simple and popular in industrial experimentations, it only provides a linear effect, if any, between the input variables and the output parameters. Furthermore, effect of interaction of any two input variables, if any, to the output parameters is not characterizable.

To compensate for the insufficiency of the screening design, it is enhanced to include center point of each input variable in experimentations. The center point of each input variable allows a quadratic effect, minimum or maximum inside explored space, between input variables and output parameters to be identifiable, if one exists. The enhancement is commonly known as response surface design to provide quadratic response model of responses. The quadratic response model can be calibrated using full factorial design (all combinations of each level of input variable) with three or more levels. However, the full factorial designs generally require more samples than necessary to accurately estimate model parameters. In

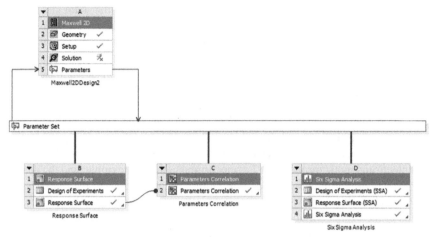

Figure 1.33   ANSYS workflow on robust design analysis based on ANSYS Maxwell as finite element analysis solution.

TABLE 1.7   IPM specifications

Performance specification	Values
DC bus	400 V
Peak torque	330 Nm
Peak power	235 kW
Max speed	15,000 rpm
Base speed	7500 rpm
Continuous power at base speed	80 kW
Continuous torque at base speed	100 Nm

TABLE 1.8   IPM dimensions

Motor design details	Values
Slots	48
Poles	8
Stator OD	220 mm
Stator axial length	150 mm
Axial laminated pack	150 mm
Stator and rotor steel	M270-35A
Permanent magnets	N38UH (Br = 1.05 T and Hc = 360 kA/m at 70°C)

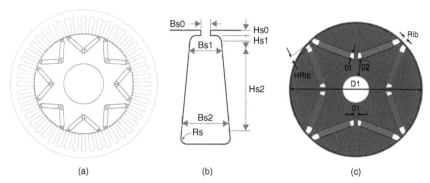

(a)   (b)   (c)

Figure 1.34   Electric motor design topology: (a) IPM 2D-cross section; (b) stator slot dimensions; (c). IPM rotor dimensions.

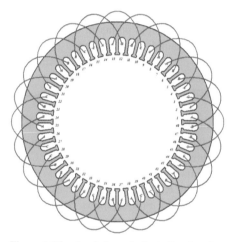

Figure 1.35   A whole-coiled winding topology.

light of the deficiency, a statistical procedure is developed to devise much more efficient experiment designs using three or five levels of each factor but not all combinations of levels, known as fractional factorial designs. In this study among these fractional factorial designs, the most popular response surface designs known as Central Composite Design (CCD) is used.

TABLE 1.9   Design parameters definition

Design variables (Parameters)	Description	Initial values
P2 (Bs0)	Stator slot opening (mm)	3
P3 (Bs2)	Top stator slot width (mm)	7
P4 (OD)	Rotor outer diameter (mm)	146
P5 (Rib)	PM bridge (mm)	1

TABLE 1.10 Design variable with lower and upper bounds

Design variables	Lower bound (LB)	Upper bound (UB)
P2 (BS0) (mm)	1	4
P3 (BS2) (mm)	6	8.5
P4 (OD) (mm)	146	148.5
P5 (RIB) (mm)	1	5

In fact, DOE is a technique used to determine the location of sampling points and is included as part of the Response Surface and Six Sigma systems. There are several versions of DOE available in engineering literature. These techniques all have one common characteristic: they try to locate the sampling points such that the space of random input parameters is explored in the most efficient way, or obtain the required information with a minimum of sampling points. Sample points in efficient locations will not only reduce the required number of sampling points, but also increase the accuracy of the response surface that is derived from the results of the sampling points. By default, the deterministic method uses a CCD, which combines one center point, the points along the axis of the input parameters, and the points determined by a fractional factorial design.

The design variables selection is presented in Table 1.10 where for each variable both lower bound and upper bound are identified. The computed minimum and maximum for selected output parameters are shown in Table 1.11. Based on the CCD, the DOE creates and solves only the design points presented in Figure 1.36.

This analysis generates the data required by Response Surface analysis. The first result of Response Surface analysis, Goodness of Fit that shows how well the Response Surface data prediction fits the expected results observed from design points is illustrated in Figure 1.37. The Goodness of Fit summarizes the difference between observed values and the values expected. In the current case study, it tests whether two samples are drawn from identical distributions.

Figure 1.38 shows local sensitivity plot for all design variables with respect to both output parameters. Important to mention is the positive sensitivity of stator slot width (P3 design variable) with respect to ripple torque (P7) and negative sensitivity with respect to average torque (P6). This analysis shows that with decreasing stator slot width the average torque increases while torque ripple decreases. This is a

TABLE 1.11 Output parameters' description

Output parameters	Calculated minimum	Calculated maximum
P6 average torque (Nm)	130.51	173.5
P7 torque ripple (%)	3.18	13.83

	A	B	C	D	E	F	G
1	Name	P2 - x1 [mm]	P3 - x2 [mm]	P4 - x4 [mm]	P5 - x5 [mm]	P6 - mean(Moving1.Torque)	P7 - ripple(Moving1.Torque)
2	1	2.5	7.25	147.25	3	151.95	6.1393
3	2	1	7.25	147.25	3	152.45	4.8273
4	3	4	7.25	147.25	3	149.69	9.3654
5	4	2.5	6	147.25	3	157.69	5.3515
6	5	2.5	8.5	147.25	3	141.86	7.5277
7	6	2.5	7.25	146	3	134.59	4.2139
8	7	2.5	7.25	148.5	3	168.12	9.7478
9	8	2.5	7.25	147.25	1	144.08	4.9175
10	9	2.5	7.25	147.25	5	155.56	6.2703
11	10	1.4437	6.3697	146.37	1.5916	139.39	3.5572
12	11	3.5563	6.3697	146.37	1.5916	136.77	4.9217
13	12	1.4437	8.1303	146.37	1.5916	131.98	4.3963
14	13	3.5563	8.1303	146.37	1.5916	130.51	5.7024
15	14	1.4437	6.3697	148.13	1.5916	164.09	5.5889
16	15	3.5563	6.3697	148.13	1.5916	163.12	7.1734
17	16	1.4437	8.1303	148.13	1.5916	150.4	7.3814
18	17	3.5563	8.1303	148.13	1.5916	149.58	9.1663
19	18	1.4437	6.3697	146.37	4.4084	145.58	3.1781
20	19	3.5563	6.3697	146.37	4.4084	142.77	6.1977
21	20	1.4437	8.1303	146.37	4.4084	137.83	4.1625
22	21	3.5563	8.1303	146.37	4.4084	136.17	7.233
23	22	1.4437	6.3697	148.13	4.4084	173.5	6.0555
24	23	3.5563	6.3697	148.13	4.4084	173.03	10.728
25	24	1.4437	8.1303	148.13	4.4084	158.72	9.4615
26	25	3.5563	8.1303	148.13	4.4084	157.77	13.832

Figure 1.36   Design of Experiments creates a reduced parametric table for further robust design analysis.

critical observation when attempting to reduce torque ripple and increase average torque in the same time. Similarly, the effect of slot opening (P2) has larger impact (positive sensitivity) on torque ripple generation than average torque production (negative sensitivity). The air-gap length controlled by outer rotor diameter (P4) shows positive sensitivity on both average torque and ripple torque predictions. This analysis tremendously helps to decide appropriate design variables for further optimization process.

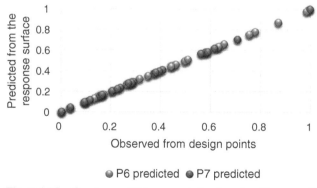

● P6 predicted ● P7 predicted

Figure 1.37   Goodness of Fit representation based on Response Surface analysis.

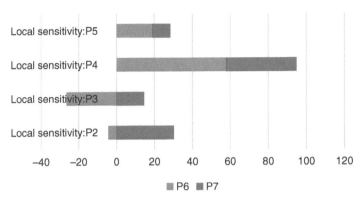

Figure 1.38   Local sensitivities for selected design variables with respect to defined output parameters.

Response Surface analysis builds 3D surface plots (Figure 1.39) illustrating average torque and ripple torque variations as function of slot opening (P2) and outer rotor diameter (P4) input design variables. Similarly, the 3D surface plots (Figure 1.40) show the surface variation of average torque and ripple torque, respectively as function of stator slot width (P3) and outer rotor diameter (P4).

At this stage of Response Surface analysis if design optimization is employed, one can further investigate the feasibility of the optimized design points with respect to imposed constraints. A screening method is used to build a trade-off analysis using 2000 evaluations to represent the feasible design points as Pareto chart or trade-off plot (Figure 1.41).

Pareto fronts search has been performed to investigate the feasible design points for further statistical studies considering tolerances. In this case study, the two defined goals are to reduce the ripple torque and increase the running torque. In such a case, one goal may be realized at the expense of the other.

The trade-off plot shows the line along which no improvement in one parameter's goal can be achieved without sacrificing the other (Pareto optimal). Pareto or non-dominated set, is a group of solutions such that selecting any one of them in place of another will always sacrifice quality for at least one objective, while improving at least one other.

Among all feasible design points, the optimized design point has been selected (Tables 1.12 and 1.13) by weighting the contribution of individual goals. In this study, ripple torque assumes higher contribution to the optimized design decision-making whereas running torque should be larger than 160 Nm.

Six Sigma Design

A Six Sigma analysis determines the extent to which uncertainties in the model affect the results of an analysis. An uncertainty (or random quantity) is a parameter whose value is impossible to determine at a given point in time (if it is time-dependent)

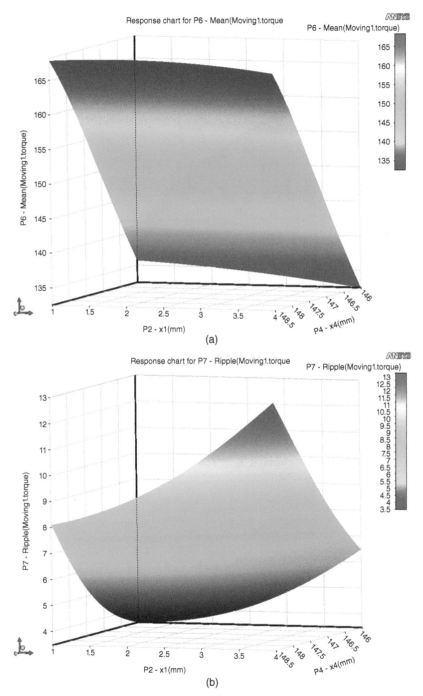

Figure 1.39   Three-dimensional surface plots: output parameters versus input design variables. (a) Average torque-P6 versus P2 and P4; (b) Ripple torque-P7 versus P2 and P4

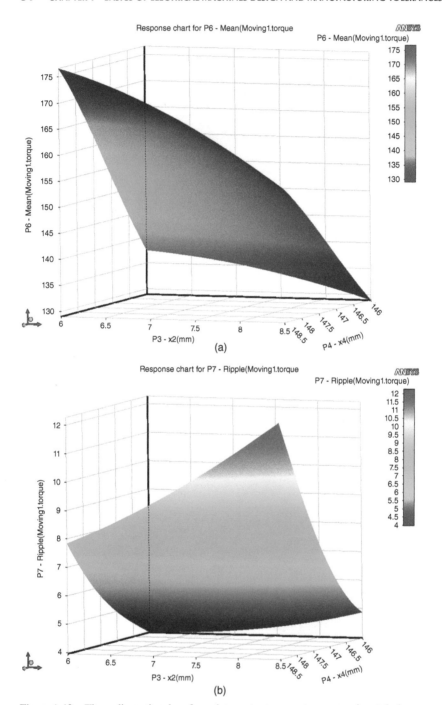

Figure 1.40 Three-dimensional surface plots: output parameters versus input design variables. (a) Average torque-P6 versus P3 and P4; (b) Ripple torque-P7 versus P3 and P4

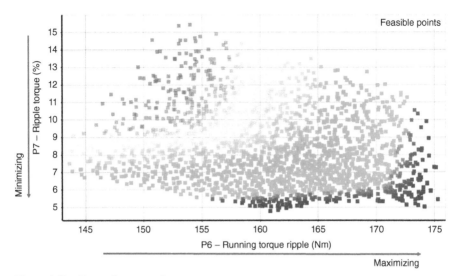

Figure 1.41   Pareto front search.

or at a given location (if it is location-dependent). There have been published some significant works on motors and drives considering Six Sigma process [9–16].

A Six Sigma analysis uses statistical distribution functions (such as the Gaussian, or normal distribution, the uniform distribution, lognormal, etc.) to describe uncertain parameters. By means of simulations one can measure numerically predicting the performance of a device or product. To assess the quality of such performance, Six Sigma analysis determines whether the design satisfies Six Sigma quality criteria. In fact, Six Sigma analysis is a statistical analysis, which when applied to a simulation generates a test of significance against the criteria of the device good quality.

TABLE 1.12   Optimized design variables

Design variables	Optimized value
P2 (BS0) (mm)	1.17
P3 (BS2) (mm)	6.08
P4 (OD) (mm)	148
P5 (RIB) (mm)	3.3

TABLE 1.13   Optimized motor design performance

Output parameters	Optimized performance
P6 average torque (Nm)	170.58
P7 torque ripple (%)	5.2673

In general, considering μ the mean of a large population with σ the population standard deviation, $\bar{x}$ the mean of a simple random sample of the population which obeys normal distribution in order to apply the central limit theorem (CLT), one can easily predict the margin errors (tolerances) estimating the population mean with certain confidence level. In this case, the confidence interval (CI) can be calculated by

$$\bar{x} \pm z^* \frac{\sigma}{\sqrt{N}}, \qquad (1.11)$$

where N is the size of sample and z^* is a parameter given by the selected confidence level.

In statistics, the very well-known *68-95-99.7* rule provides with the confidence levels in terms of probabilities of values that lie within a band around the mean in a normal distribution with a width of two, four, or six standard deviations. The data outside of these intervals do not obey the quality criteria, in other words, for such design variations the device or product fails to sustain the level of desired performances. For example, if the confidence level of the desired quality is 99.7% then only $1-0.997 = 0.003$ or almost 3000 parts out of every 1 million manufactured fail. Similarly, a product has Six Sigma quality—with the confidence level as high as 99.99966%—if only 3.4 parts out of every 1 million manufactured fail. This quality definition is based on the assumption that an output parameter relevant to the quality and performance assessment follows a Gaussian (normal) distribution, as shown in Figure 1.42.

The main goal of this study is to assess the design fitness based on Six Sigma analysis on selected electric motor topology. Considering the level of confidence as

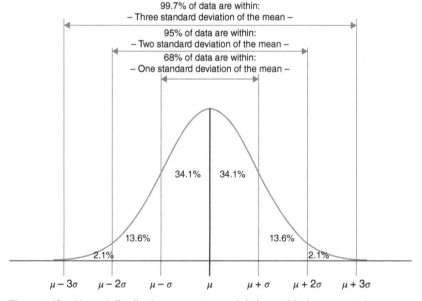

Figure 1.42   Normal distribution parameters and their graphical representation.

99.7%, the confidence interval is $\mu \pm 3\sigma$. Assuming the mean of population as the mean of individual simulations (sample) for each design variable, one can easily calculate the individual simulation standard deviation as

$$\mu \pm 3\sigma = \mu \pm \Delta x \cdot \mu \qquad (1.12)$$

where Δx could be interpreted as the tolerance value for corresponding design variable. Then one can assume

$$\sigma_x = \frac{\Delta x \cdot \mu}{3} \qquad (1.13)$$

where σ_x represents the standard deviation of individual samples.

In this study, the optimized IPM design topology is selected according to Response Surface analysis. Based on this assumption, the design variables corresponding to P2, P3, P4, and P5 are selected as baseline design variables that provide with high torque profile and low ripple torque characteristic. Statistically, the values of the Response Surface output parameters are considered to be the parameters of the population. In such condition, these values are used to test the statistical significance when Six Sigma analysis is performed. Although the design variable P4 that controls air-gap length is set fixed at 148 mm providing with an air-gap length of 1 mm, all other design variables are introduced assuming their nominal values obey normal distribution with mean values corresponding to the optimized design point and standard deviations calculated based on tolerances as described in (1.13) and shown in Table 1.14. Considering the tolerances of design variables as 3σ, one can calculate the simulated design variables' standard deviations at given uncertainty 10%, 5%, and 1%, respectively. The uncertainty of design variable can change the performance of the electric motor. The main goal is to use such design variable distribution to estimate the variation band of the output parameters according to the uncertainty.

Based on this table, Six Sigma can combine different standard deviations for individual design variables to control the variation band of the output parameters. An iterative procedure might be applied using Six Sigma analysis to increase the test of significance ensuring high quality design and in the same time to balance the cost of generating such quality. At each iteration, these values are selected based on sensitivity analysis which shows which design variable variation has more effect on selected output parameters. The importance of such analysis relates very well to the manufacturing tolerance process [17] providing with the level of confidence if the electric motor will run at the targeted performance.

TABLE 1.14 Design variable tolerances

Design variables	Mean value	STD. deviation at 10% uncertainty	STD. deviation at 5% uncertainty	STD. deviation at 1% uncertainty
P2 (BS0) (mm)	1.17	0.039	0.0195	0.0039
P3 (BS2) (mm)	6.08	0.203	0.101	0.0203
P5 (RIB) (mm)	3.3	0.11	0.055	0.011

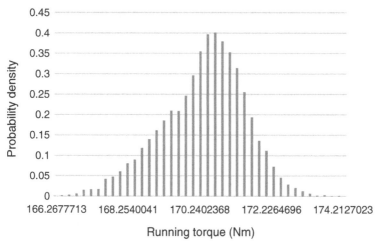

Running torque (Nm)

Figure 1.43   Running torque distribution on 10%-10%-10% design variable tolerances.

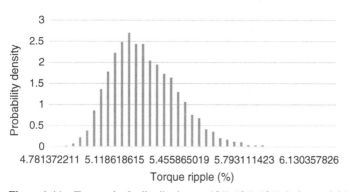

Torque ripple (%)

Figure 1.44   Torque ripple distribution on 10%-10%-10% design variable tolerances.

If Six Sigma analysis is performed with a tolerance of 10% for all design variables (P2, P3, P5) based on 3σ level (Table 1.15), the variation of individual output parameters as statistical distribution is shown in Figures 1.43 and 1.44.

The distributions of running torque (which is the average torque of electric motor at steady state) and torque ripple show following statistics:

TABLE 1.15   10%-10%-10% tolerances on design variables

10% tolerance on P2 10% tolerance on P3 10% tolerance on P5	Mean value	Probability value
Running torque (Nm)	170.24	0.4874 (P > 170.58 Nm)
Ripple torque (%)	5.456	0.6025 (P < 5.2673%)

In Table 1.15, "mean value" represents the mean for Six Sigma simulation whereas "probability value" represents the probability which is a hypothesis testing procedure based on the z-statistic. In this case study for given hypothesized population mean, this is the probability that the sample mean would be greater or less than the average of simulations statistically seen as observed data. In this context, the probability will measure the acceptance levels that the value corresponding to running torque will be greater than the optimized value of 170.58 Nm and the value corresponding to torque ripple will be less than the optimized value of 5.2673%. As an important note in these results there is no significant skewness (mean value practically is equal to median value). This observation is valid for entire study presented in this section.

The probability value can be also evaluated using z-test computed as

$$z^* = \frac{\bar{x} - \mu}{\sigma / \sqrt{N}} \tag{1.14}$$

Using (1.14), one can calculate the probabilities of higher or smaller values using the standard normal distribution table as a $P(z > z^*)$ for running torque and $P(z < z^*)$ for torque ripple.

According to this analysis, almost 60.25% of all designs will have torque ripple values less than 5.2673%. Similarly, almost 48.74% of all designs will have running torque values greater than 170.58 Nm. Since the running torque level of confidence could use higher error margins, the main goal is to provide with higher sigma-level confidence on ripple torque prediction.

In order to reduce scatter of the outputs, selection of design variables requires smaller tolerances. To select the appropriate design variable, sensitivity analysis is performed as shown in Figure 1.45.

This analysis shows that the design variable related to slot width (P3) has major impact on both running torque and torque ripple. Basically, P3 design variable

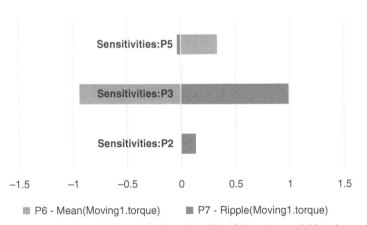

Figure 1.45   Sensitivity results for 10%-10%-10% design variable tolerances.

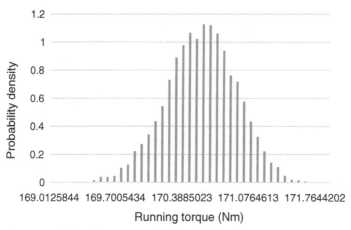

Figure 1.46   Running torque distribution on 10%-10%-1% design variable tolerances.

exercises opposite sensitivities on running torque. Therefore, the shown positive sensitivity illustrates that any decrease on slot width will generate a decrease in the ripple torque value whereas highlighted negative sensitivity demonstrates that any decrease on slot width will generate an increase in running torque value. This is an excellent example to show sensitivity analysis when a design variable needs to be selected for further optimization. In this study, practically a reduction of the slot width value will increase the torque and reduce the torque ripple. Therefore, a lower tolerance (1%) is considered for design variable P3.

Six Sigma analysis is then performed with a tolerance of 10% for design variables P2 and P5 and 1% tolerance for design variable P3 based on same 3σ level. Figures 1.46 and 1.47 illustrate the new distribution of running torque and torque ripple characteristics.

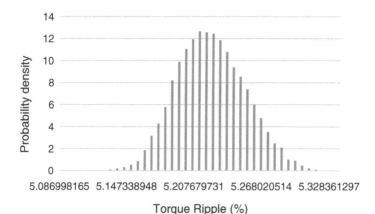

Figure 1.47   Torque ripple distribution on 10%-10%-1% design variable tolerances.

TABLE 1.16   10%-10%-1% tolerances on design variables

10% tolerance on P2 10% tolerance on P5 1% tolerance on P3	Mean value	Probability value
Running torque (Nm)	170.39	0.4667 (P > 170.58 Nm)
Ripple torque (%)	5.207	0.9315 (P < 5.2673%)

These plots show a significant reduction of scatter of both distributions. Table 1.16 presents the new mean value and probability results as numerical measurement to accept the statistical hypothesis for both running torque and torque ripple characteristics.

According to this analysis, almost 46.67% of all designs will have running torque values greater than 170.58 Nm. As far ripple torque characteristic is concerned the statistical test shows better level of significance. Almost 93.15% of all designs will have torque ripple values less than 5.2673%.

The scatter of outputs centralizes around mean of the population defining tighter tolerances for most sensitive parameters. Sensitivity analysis illustrated in Figure 1.48 shows further improvements if P5 (PM bridge) and P2 (slot opening) are considered with lower tolerance. The results of new Six Sigma analysis are presented in Table 1.17.

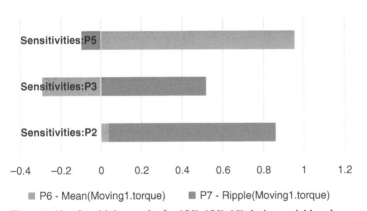

■ P6 - Mean(Moving1.torque)   ■ P7 - Ripple(Moving1.torque)

Figure 1.48   Sensitivity results for 10%-10%-1% design variable tolerances.

TABLE 1.17   5%-5%-1% tolerances on design variables

5% tolerance on P2 5% tolerance on P5 1% tolerance on P3	Mean value	Probability value
Running torque (Nm)	170.51	0.4411 (P > 170.58 Nm)
Ripple torque (%)	5.212	94.81 (P < 5.2673%)

TABLE 1.18   5%-1%-1% tolerances on design variables

5% tolerance on P2 1% tolerance on P5 1% tolerance on P3	Mean value	Probability value
Running torque (Nm)	170.54	0.3897 (P > 170.58 Nm)
Ripple torque (%)	5.22	0.9942 (P < 5.2673%)

Depending on targeted application, such analysis should always be a trade-off between the quality of the design and the cost of manufacturing to maintain low tolerance values. From last Six Sigma simulation (as reported in Table 1.18), tightening the tolerances on PM bridge dimension (P5) increases the probability of getting less torque ripples with a trade-off on getting higher running torque with respect to previous Six Sigma analysis.

REFERENCES

[1] D. M. Ionel and M. Rosu, "Short course: Introduction to electric machine design for manufacturing," in *IEEE ITEC 2014*, Dearborn, MI, Jun. 2014.

[2] D. M. Ionel, S. J. Dellinger, R. J. Heideman, and A. E. Lesak, "Stator assembly for an electric machine and method of manufacturing the same," U.S. Patent 7348706, Mar. 2008.

[3] D. M. Ionel, "High-efficiency variable-speed electric motor drive technologies for energy savings in the US residential sector," in *Rec. IEEE Optim. 2010 Conf.*, Brasov, RO, May 2010, pp. 1403–1414.

[4] A. Fatemi, D. M. Ionel, N. A. O. Demerdash, S. Stretz, and T. M. Jahns, "RSM-DE-ANN sensitivity analysis of material cost in PM motors with distributed and concentrated windings," in *Rec. IEEE ECCE 2016 Congress*, Milwaukee, WI, Sep. 2016, p. 7.

[5] M. Popescu, I. Foley, D. A. Staton, and J. E. Goss, "Multi-physics analysis of a high torque density motor for electric racing cars," in *Rec. IEEE ECCE 2015 Congress*, Montreal, QC, Sep. 2016, p. 8.

[6] A. Fatemi, D. M. Ionel, M. Popescu, and N. A. O. Demerdash, "Design optimization of spoke-type PM motors for formula E racing cars," in *Rec. IEEE ECCE 2016 Congress*, Milwaukee, WI, Sep. 2016, p. 8.

[7] J. Pyrhonen, T. Jokinen, and V. Hrabovcova, Design of Rotating Electrical Machines. John Wiley & Sons, 2010.

[8] J. Hendershot and T. J. E Miller, Design of Brushless Permanent Magnet Machines. Motor Design Books LLC, 2010.

[9] J.-S. Hwang, K.-H. Kim, Y.-T. Kim, and S.-M. Baek, "Parameter optimization of field oriented control with 6 sigma tool," *IEEE ISIE*, vol. 3, pp. 1866–1870, Jun. 2001.

[10] G. Ombach and J. Junak, "Design of PM brushless motor taking into account tolerances of mass production - six sigma design method," *IAS Annual Meeting*, vol. 49, no. 7, pp. 2139–2146, Sep. 2007.

[11] X. Meng, S. Wang, J. Qiu, Q. Zhang, J. G. Zhu, Y. Guo, and D. Liu,"Robust multilevel optimization of PMSM using design for six sigma," *IEEE Trans. Magn.*, vol. 47, no. 10, pp. 3248–3251, Oct. 2011.

[12] G. Lei, Y. G. Guo, J. G. Zhu, T. S. Wang, X. M. Chen, and K. R. Shao, "System level six sigma robust optimization of a drive system with PM transverse flux machine," *IEEE Trans. Magn.*, vol. 48, no. 2, pp. 923–926, Feb. 2012.

[13] G. Lei, J. G. Zhu, Y. G. Guo, J. F. Hu, W. Xu, and K. R. Shao, "Robust design optimization of PM-SMC motors for six sigma quality manufacturing," *IEEE Trans. Magn.*, vol. 49, no. 7, pp. 3953–3956, Jul. 2013.

[14] G. Lei, J. Zhu, Y. Guo, K. Shao, and W. Xu, "Multiobjective sequential design optimization of PM-SMC motors for six sigma quality manufacturing," *IEEE Trans. Magn.*, vol. 50, no. 2, Feb. 2014.

[15] G. Lei, J. G. Zhu, Y. G. Guo, J. F. Hu, W. Xu, and K. R. Shao, "Robust design optimization of PM-SMC motors for six sigma quality manufacturing," *IEEE Trans. Magn.*, vol. 49, no. 7, pp. 3953–3956, Jul. 2013.

[16] G. Lei, T. Wang, J. Zhu, Y. Guo, and S. Wang, "System-level design optimization method for electrical drive systems—Robust approach," *IEEE Trans. Ind. Electron.*, vol. 62, no. 8, pp. 4702–4713, Aug. 2015.

[17] Y.-K. Kim, J.-P. Hong, and J. Hur, "Torque characteristic analysis considering the manufacturing tolerance for electric machine by stochastic response surface method," *IEEE Trans. Ind. Appl.*, vol. 39, no. 3, pp. 713–719, May/Jun. 2003.

CHAPTER 2

FEM-BASED ANALYSIS TECHNIQUES FOR ELECTRICAL MACHINE DESIGN

2.1 T–Ω FORMULATION

The **T–Ω** formulation is a very efficient formulation for solving eddy-current problems. The scalar potential Ω is represented by nodal shape functions in the entire domain while the current vector potential **T** is represented by edge-based shape functions restricted to only conducting regions.

To be more general, the magnetic field **H** can be further split into four components and expressed as

$$\mathbf{H} = \mathbf{H}_p + \mathbf{T} + \nabla\Omega + \sum i_k \mathbf{H}_k, \tag{2.1}$$

where $\mathbf{H}_p$ is the source field component associated with all known current excitations, i_k is the unknown current in the voltage-driven coil k, and $\mathbf{H}_k$ is the field component created by 1 A current flowing in the voltage-driven coil k and zero currents in all others [1]. In equation (2.1), $\mathbf{H}_p$ is precalculated and serves as a particular solution of $\nabla \times \mathbf{H}_p = \mathbf{J}$. Unknown currents are to be determined by coupling with voltage balance equations for all voltage-driven coils. A coil can be either a solid conductor with induced eddy currents or stranded wires with uniformly distributed current density.

2.1.1 Basic Field Equations and Norton Nonlinear Iteration Form

The basic field equations are given as

$$\begin{cases} \nabla \times \dfrac{1}{\sigma}\nabla \times \mathbf{H} + \dfrac{d\mathbf{B}}{dt} = 0 \\[2mm] \nabla \cdot \mathbf{B} = 0 \end{cases} \tag{2.2}$$

Multiphysics Simulation by Design for Electrical Machines, Power Electronics, and Drives, First Edition.
Marius Rosu, Ping Zhou, Dingsheng Lin, Dan Ionel, Mircea Popescu, Frede Blaabjerg, Vandana Rallabandi, and David Staton.

For the sake of conciseness without loss of generality, the following derivation in this section will not involve voltage excitation, circuit coupling, and rigid motion. With transient T–Ω formulation, let us define

$$\begin{cases} \mathbf{F}_1\left(\Omega, \mathbf{T}\right) = \nabla \times \dfrac{1}{\sigma}\nabla \times \mathbf{H} + \dfrac{d\mathbf{B}}{dt} \\ F_2\left(\Omega, \mathbf{T}\right) = \nabla \cdot \mathbf{B} \end{cases} \tag{2.3}$$

The corresponding Newton iteration forms are

$$\begin{cases} -\mathbf{F}'^{k}_{1(\partial \mathbf{T})}\left(\mathbf{T}^{k+1} - \mathbf{T}^k\right) - \mathbf{F}'^{k}_{1(\partial \Omega)}\left(\Omega^{k+1} - \Omega^k\right) = \mathbf{F}^k_1 \\ -F'^{k}_{2(\partial \mathbf{T})}\left(\mathbf{T}^{k+1} - \mathbf{T}^{\mathbf{k}}\right) - F'^{k}_{2(\partial \Omega)}\left(\Omega^{k+1} - \Omega^k\right) = F^k_2 \end{cases} \tag{2.4}$$

Taking the Frechet partial derivative of equation (2.3) yields

$$\mathbf{F}'_{1(\partial \mathbf{T})} = \nabla \times \frac{1}{\sigma}\nabla \times () + \frac{d}{dt}\left([\hat{\mu}]()\right) \tag{2.5}$$

$$\mathbf{F}'_{1(\partial \Omega)} = \frac{d}{dt}\left([\hat{\mu}]\nabla()\right) \tag{2.6}$$

$$F'_{2(\partial \mathbf{T})} = \nabla \cdot [\hat{\mu}](), \tag{2.7}$$

$$F'_{2(\partial \Omega)} = \nabla \cdot [\hat{\mu}]\nabla() \tag{2.8}$$

where tensor $[\hat{\mu}]$ can be considered as an equivalent dynamic permeability and defined by

$$[\hat{\mu}] = \frac{\partial \mathbf{B}}{\partial \mathbf{H}} = \begin{bmatrix} \partial B_x/\partial H_x & \partial B_x/\partial H_y & \partial B_x/\partial H_z \\ \partial B_y/\partial H_x & \partial B_y/\partial H_y & \partial B_y/\partial H_z \\ \partial B_z/\partial H_x & \partial B_z/\partial H_y & \partial B_z/\partial H_z \end{bmatrix} \tag{2.9}$$

In the case of isotropic material, $[\hat{\mu}]$ is expressed as [2]

$$[\hat{\mu}] = \begin{bmatrix} \mu & 0 & 0 \\ 0 & \mu & 0 \\ 0 & 0 & \mu \end{bmatrix} + (\mu' - \mu) \cdot \begin{bmatrix} c_x^2 & c_x c_y & c_x c_z \\ c_y c_x & c_y^2 & c_y c_z \\ c_z c_x & c_z c_y & c_z^2 \end{bmatrix} \tag{2.10}$$

where $\mu = B/H$, $\mu' = dB/dH$, and

$$\begin{cases} c_x = H_x/H \\ c_y = H_y/H \\ c_z = H_z/H \end{cases} \tag{2.11}$$

with

$$\begin{cases} B = \sqrt{B_x^2 + B_y^2 + B_z^2} \\ H = \sqrt{H_x^2 + H_y^2 + H_z^2} \end{cases} \tag{2.12}$$

It can be seen from equation (2.10) that the tensor $[\hat{\mu}]$ has off-diagonal entries. It is these off-diagonal elements that model the physics of the cross effects of magnetic saturation.

Next, let Ω and $\mathbf{T}$ be approximated as

$$\begin{cases} \Omega = \tilde{\alpha}\underline{\Omega} \\ \mathbf{T} = \tilde{\mathbf{t}}\underline{T} \end{cases}, \tag{2.13}$$

where $\tilde{\alpha} = [\alpha_1, \alpha_2, ..., \alpha_n]$ and $\tilde{\mathbf{t}} = [\mathbf{t}_1, \mathbf{t}_2, ..., \mathbf{t}_m]$ are scalar and vector basis functions, respectively, defined locally over each element. $\underline{\Omega} = [\Omega_1, \Omega_2, ..., \Omega_n]^T$ are the values of Ω at the mesh nodes and $\underline{T} = [T_1, T_2, ..., T_m]^T$ are the values of $\mathbf{T}$ projected at the mesh edges. Thus, the equation (2.4) becomes

$$\mathbf{F}'^k_{1(\partial \mathbf{T})} \, \tilde{\mathbf{t}} \left(\underline{T}^{k+1} - \underline{T}^k \right) + \mathbf{F}'^k_{1(\partial \Omega)} \, \tilde{\alpha} \left(\underline{\Omega}^{k+1} - \underline{\Omega}^k \right) = -\mathbf{F}^k_1, \tag{2.14}$$

$$F'^k_{2(\partial \mathbf{T})} \, \tilde{\mathbf{t}} \left(\underline{T}^{k+1} - \underline{T}^k \right) + F'^k_{2(\partial \Omega)} \, \tilde{\alpha} \left(\underline{\Omega}^{k+1} - \underline{\Omega}^k \right) = -F^k_2 \tag{2.15}$$

By applying the Galerkin method and multiplying both sides of equation (2.14) by $\underline{\mathbf{t}}$ and equation (2.15) by $\underline{\alpha}$, and then integrating over the problem domain V, we get

$$\begin{cases} J_{11} \left(\underline{T}^{k+1} - \underline{T}^k \right) + J_{12} \left(\underline{\Omega}^{k+1} - \underline{\Omega}^k \right) = \underline{R}_1 \\ J_{21} \left(\underline{T}^{k+1} - \underline{T}^k \right) + J_{22} \left(\underline{\Omega}^{k+1} - \underline{\Omega}^k \right) = \underline{R}_2 \end{cases} \tag{2.16}$$

where $J_{11}, J_{12}, J_{21},$ and J_{22} are the blocks of the Jacobian matrix

$$\begin{cases} J_{11} = \int \underline{\mathbf{t}} \cdot \mathbf{F}'^k_{1(\partial \mathbf{T})} \tilde{\mathbf{t}} dV \\[2mm] J_{12} = \int \underline{\mathbf{t}} \cdot \mathbf{F}'^k_{1(\partial \Omega)} \tilde{\alpha} \, dV \\[2mm] J_{21} = \int \underline{\alpha} F'^k_{2(\partial \mathbf{T})} \tilde{\mathbf{t}} \, dV \\[2mm] J_{22} = \int \underline{\alpha} F'^k_{2(\partial \Omega)} \tilde{\alpha} \, dV \end{cases} \tag{2.17}$$

and $\underline{R}_1$ and $\underline{R}_2$ are the residual vectors

$$\begin{cases} \underline{R}_1 = -\int \underline{\mathbf{t}} \cdot \mathbf{F}^k_1 \, dV \\[2mm] \underline{R}_2 = -\int \underline{\alpha} \, F^k_2 \, dV \end{cases} \tag{2.18}$$

To consider the material property variations within the element, Gauss quadrature numerical integration is adopted to evaluate the integrals in equations (2.17) and (2.18).

2.1.2 Strategies for Accelerating Nonlinear Convergence

The Newton–Raphson technique for the solution of nonlinear problems in finite element analysis (FEA) has been widely accepted due to its quadratic convergence characteristics [3]. However, for highly nonlinear problems, the Newton sequence with an arbitrary initial guess may converge at a very slow rate, or oscillate, or even diverge. It is also known that convergence is more difficult with the magnetic scalar potential than with magnetic vector potential [4]. Hence, under-relaxation is commonly used to improve convergence. If an optimum relaxation factor, which minimizes the total square of the 2-norm of the residual obtained from the finite element discretization is introduced at each step of nonlinear iteration, a convergent solution can always be obtained using a linear search algorithm [5, 6]. However, it may take a very long time to find optimum relaxation factor because a large number of repeated evaluations for the residual are required.

In addition, it is understandable that the use of the optimum relaxation factor does not guarantee fast convergence and good efficiency because the "optimum" is only in the global sense and thus may not be suitable for some of the elements. This may lead to poor local convergence due to large overshoot correction or possible oscillation. It was observed that even with only a few such "bad" elements, there was a significant impact on the convergence rate of the Newton sequence, particularly in the case of coarse mesh and large field gradient in nonlinear materials. This undesirable local convergence behavior may also have a significant impact on solution accuracy, such as in the case of considering induced eddy current in the nonlinear region. In such a case, a poorly converged solution in those "bad" elements can directly produce unphysical eddy current, and during a transient analysis, the numerical errors will propagate and accumulate with time.

To this end, a scheme was proposed to efficiently find the optimum relaxation factor for improving global convergence performance. Further, to address the local convergence issue, a local damping scheme is introduced [7]. This scheme damps the updating of the nonlinear material property for a small portion of elements that exhibit the largest changes in the equivalent dynamic permeability.

The nonlinearity in the obtained system of equation (2.16) arises from the fact that the entries of the obtained global Jacobian matrix and the residual vector are functions of the permeability or the equivalent dynamic permeability assigned to each nonlinear element. These elemental permeabilities are functions of the unknowns to be solved. Therefore, an iterative process is necessary. Figure 2.1 shows the flowchart of the Newton–Raphson iteration scheme. When the relaxation factor α is equal to unity, Figure 2.1 represents the ordinary Newton–Raphson method. Under-relaxation is commonly used to achieve convergence or to improve the convergence rate. As the nonlinear iteration converges, the residual for each unknown should approach zero. Therefore, the optimal under-relaxation factor is the one which minimizes the square of the 2-norm of the residual $(\|R\|_2)^2$ with each iteration.

The search for the optimal relaxation factor would normally require significant computation time since large number of repeated evaluations of the residual function may be required. In order to reduce computation time, the following searching

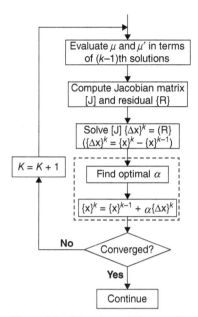

Figure 2.1   Flowchart of Newton–Raphson iteration sequence with the use of under-relaxation factor.

algorithm for the optimal under-relaxation factor is proposed: let the relaxation factor α (between 1 and 0) be equally sampled with step of 0.2 as shown in Figure 2.2. Here, $\alpha = 0$ corresponds to the previous Newton iteration solution and $\alpha = 1$ corresponds to the ordinary nonlinear Newton iteration scheme. The residuals at these two values of α are already available as part of the regular solution process. Thus, the evaluation of the residual functions can start from $\alpha = 0.8$. For each sampled value of α, after the solution candidate is obtained, we compute the H field, update the permeability, and compute the residual. If the current residual is smaller than the previous residual, we continue to the next value of α. If the current residual is greater, we use the current and the previous two values of α to construct a quadratic polynomial and find the optimal under-relaxation factor associated with the minimum residual value. A special case occurs when the residual at $\alpha = 0.8$ is greater than that at $\alpha = 1$. In such a case, α_{opt} can be simply chosen as 1. If for $\alpha = 0.2$, the corresponding computed residual value still does not increase, the sub-region of α between 0.2 and 0 is further sub-divided with a step of 0.05 and the search is continued.

In order to improve efficiency, a large step size of 0.2 is used in the above algorithm. This has the added advantage of making the algorithm less sensitive to local minimum. However, the large step size may lead to a large error in the computed optimal under-relaxation factor. To address this, we can insert two additional α points in between the above obtained three points and compute corresponding residuals. This will effectively reduce the step size to 0.1 and the optimal under-relaxation factor will be determined based on the appropriately selected set of three points: the point with the smallest residual and two neighboring points on each side.

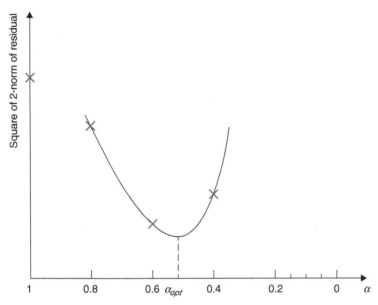

Figure 2.2   Search for the optimal under-relaxation factor.

As mentioned above, the use of the optimum relaxation factor does not necessarily guarantee fast convergence and good efficiency because the "optimum" is measured only in the global sense and some elements may still have poor local convergence. To improve convergence for such cases, a local damping factor β is introduced to damp the updating of the equivalent dynamic permeability tensor in the process of computing the Jacobian matrix

$$[\hat{\mu}]^{k+1}_{\text{actual}} = [\hat{\mu}]^k + \beta \left([\hat{\mu}]^{k+1} - [\hat{\mu}]^k \right) \tag{2.19}$$

It should be emphasized that this modification should be applied to only a very small percentage of elements (less than 1% of total nonlinear elements) with the highest rate of change in the equivalent dynamic permeability tensor. The small percentage ensures that the convergence rate and the efficiency of Newton–Raphson approach will not be adversely affected. In our investigation with many test cases, we have arrived at the following empirical formula for determining n, the number of elements to be damped

$$n = \begin{cases} 3 \cdot \ln \left(N_e/n_0 \right) & \text{when } N_e > n_0 \\ 0 & \text{when } N_e \leq n_0 \end{cases}, \tag{2.20}$$

where N_e is the total number of nonlinear elements in the solved domain and n_0 is the nonlinear element size threshold. The local damping algorithm is only applied if the element size is greater than this threshold. Our investigation has shown that $n_0 = 500$ is a reasonable choice.

For the local damping process, the first step is to identify the n elements with the highest change rate k_e out of the entire set of nonlinear elements, where k_e is determined by

$$k_e = \max\left(\frac{\left|\hat{\mu}_{11}^{k+1} - \hat{\mu}_{11k}\right|}{\hat{\mu}_{11k}}, \frac{\left|\hat{\mu}_{22}^{k+1} - \hat{\mu}_{22k}\right|}{\hat{\mu}_{22k}}, \frac{\left|\hat{\mu}_{33}^{k+1} - \hat{\mu}_{33k}\right|}{\hat{\mu}_{33k}}\right) \tag{2.21}$$

Next, the smallest k_e in the list is considered to be the reference damping rate k_{ref}. The ratio of the reference rate to the actual rate for each of the n elements is then computed as

$$k_\mu = \frac{k_{\text{ref}}}{k_e} \tag{2.22}$$

Finally, the damping factor for each of the n elements is determined by

$$\beta = 0.35 + 1.5e^{-1/k_\mu} \tag{2.23}$$

The value of the local damping factor computed from equation (2.23) is approximately between 0.35 and 0.9. The lower bound is used to avoid over damping for k_μ less than 0.35; the upper bound is applied to prevent altering of the convergence property of the Newton–Raphson method due to a trivial modification. In fact, for any element with $k_\mu \geq 0.9$, the local damping step can be simply skipped.

2.1.3   Challenges—Multiply Connected Domains

As discussed above, **T**--Ω formulation is a very efficient formulation for solving eddy-current problems. But when the conducting regions contain holes, the non-conducting region becomes multiply connected and the scalar potential may become multivalued. The topic of treating multiply connected regions has drawn considerable attention [8–10]. One way is to fill the holes by fake conductors of very low conductivity. Another approach is to create a surface cut with a potential jump equal to the total current in the coil [8]. It is also possible to treat multiply connected regions based on manually identifying each conduction path for every hole and terminal [9, 11]. More efficiently, the tree-co-tree technique can be applied to make multiply connected regions simply connected for both the source field computation [12] and the eddy-current computation [10, 13] (More rigorously, the requirement of domains being simply connected can be relaxed to being loop-free because simply connected regions are loop-free, but not all loop-free regions are simply connected [14]. Here, we still use the common phrase "simply connected".)

When 3D transient solutions take rigid motion into account, new challenge is presented in handling multiply connected regions. This is due to the use of non-conforming meshes in coupling two independent meshes at each time step. Another difficulty is related to the support of periodic boundary conditions. In [15], the automatic cut generation algorithm described in [14] is generalized for 3D transient solutions including rigid motion.

Concept of Cutting Domains

In our study, domain R contains a conducting region R_C with eddy currents and an excitation coil region R_S. Normally, both conducting region R_C and excitation coil region R_S contain holes. For the sake of simplicity, let $\mathbf{H}_S$ represent excitation fields $\mathbf{H}_p$ and $\mathbf{H}_k$ as expressed in equation (2.1). $\mathbf{H}_S$ has to be constructed in a simply connected region denoted as $R_{S\Sigma}$ which is composed of R_S and the cutting domain $R_{\Sigma 0}$. $R_{\Sigma 0}$ is one layer of elements filling all of the excitation coil holes so that the boundary condition $\mathbf{n} \times \mathbf{H}_S = 0$ can be imposed on $\partial R_{S\Sigma}$ without violating Ampere's law. In the cutting domain $R_{\Sigma 0}$, the condition $\nabla \times \mathbf{H}_S = 0$ is strongly imposed by setting the line integral of $\mathbf{H}_S$ to zero on the tree edges and assigning to each co-tree edge the total current enclosed by the loop generated by this co-tree edge together with corresponding tree edges. All the loop integrals of $\mathbf{H}_S$ associated with every co-tree edge outside $\partial R_{S\Sigma}$ are zero. Thus, an efficient cutting domain is the one that forces $\mathbf{H}_S$ to be zero on as many co-tree edges as possible.

Similarly, the current vector potential $\mathbf{T}$ has to be applied in a simply connected region $R_{C\Sigma}$ that includes both the conducting region R_C and the cutting domains $R_{\Sigma i}$, $i = 1, 2, ..., n$, where n is the number of holes. The current vector potential $\mathbf{T}$ on the cutting domains has the property of modeling a zero *curl* field that cannot be expressed as the gradient of the scalar potential Ω. To assure a single-valued scalar potential, the line integral of $\mathbf{T}$ along an edge inside $R_{\Sigma i}$, which is part of a non-contractible loop, should be equal to the induced eddy current enclosed by the loop. There is one degree of freedom of $\mathbf{T}$ for each cutting domain $R_{\Sigma i}$. If any conductor happens to be an excitation coil, the corresponding degree of freedom must be set to zero because the source field $\mathbf{H}_S$ has satisfied the Ampere's law and $\mathbf{T}$ only redistributes the current. In such special cases, the generation of cuts can be avoided [10].

Automatic Cutting Domain Creation

The automatic generation of cuts is based on the automatic identification of edges for every element. If an edge is the third side of a triangle whose other two sides have been identified, this third edge will be identified and close the three-edge loop of the triangle. Otherwise, there are two cases to consider. If there is a path of previously identified edges connecting one end of the edge to the other, this edge cannot be identified since this would form a closed path which might violate Ampere's law. On the other hand, if there is no such connecting path, the edge can be safely identified. If all three edges of a triangle have been identified, the triangle becomes single connected. This identification process can be implemented using the scheme below [14].

Start by giving every mesh node a label with the initial value "0," and also introduce a loop index initialized to zero (loop_index = 0). Then go through each triangle according to the following rules:

a. If both nodes of an edge are labeled "0," the edge can be identified and a new loop starts with loop_index = loop_index+1 and both nodes are labeled with the index of the loop;

b. If one node is labeled "0" and the other is not, give the node labeled "0" the same label as the other node and the edge is identified;

c. If two nodes have different labels but neither of them is "0," the two nodes have been in two different loops. The two loops can be connected together by relabeling all nodes on loop 2 with the same label on node 1 and the edge will be identified;

d. If two nodes have the same label, but not "0" and the edge has not been identified, then the edge can only be identified if the other two edges of the triangle have been identified.

Go through the triangle list repeatedly using the rules (a)–(d) until there is no additional edge to be identified when going through the full triangle list.

Now let us look at how the above scheme is applied to the automatic generation of cutting domains. For the sake of clarity, the following discussion takes current vector potential **T** as the example. In a bounded region R that is the union of conducting region R_C and non-conducting region R_n, we search for a maximum set of tetrahedrons R_m in R_n such that the set R_m is simply connected instead of detecting the holes and looking directly for the cutting domains. By saying maximum, we mean that the domain R_m becomes multiply connected by adding any additional tetrahedron to the set R_m. Once such a set is determined, the remaining tetrahedrons in R_n form the cutting domain R_Σ ($=R_n \backslash R_m$). Such a cutting domain is not unique but works for our purpose. The procedure for generating cutting domains consists of two steps.

Step 1. Make surface cuts on the surface of conducting regions. Scan all triangles on the conductor surfaces and for each triangle examine whether each edge can be identified or not according to the rules above. A triangle with all three edges identified is added to the set of singly connected surfaces. This scanning process is repeated until there are no more triangles to be added. The rest of the triangles that do not belong to the set of singly connected surfaces create surface cuts Σ_i on the conductor surfaces.

Step 2. Extend the surface cuts to the non-conducting region. Scan all the tetrahedrons in the non-conducting region R_n. Start from tetrahedrons with a singly connected triangle on the conductor surfaces. For each tetrahedron, examine each triangle by identifying the edges according to the rules above. If all four triangle faces are singly connected, we add this tetrahedron to the set of singly connected domain. For the selection order of the tetrahedrons, select with priority from tetrahedrons with one or more neighboring tetrahedrons having been included in the set of singly connected domain. This scanning process is repeated over and over until no more tetrahedrons are to be added. Finally, the remaining tetrahedrons that are not included in the singly connected domain form the desired cutting domains R_Σ.

To find cutting domains in periodic multiply connected domains with matching (master and slave) boundaries, we need to ensure that the trace of cutting domains on a slave boundary is the same as the traces on the corresponding master boundary. A convenient and reliable approach is: whenever an edge on master or slave boundary is identified, the corresponding slave or master edge is also marked as identified by assigning the same index number to the nodes of the edge. Similarly, both master and slave edges are also marked at the same time with the identical label and edge value with appropriate sign depending on full periodic or anti-periodic boundary condition.

The same scheme is also applied to the identification of triangles on matching boundaries. As a result, the connectivity and identified traces of cutting edges on the slave boundary are always identical to the one on the master boundary.

Mesh Coupling Due to Motion

When motion is involved, two independent sets of surface meshes must be coupled together after an arbitrary displacement of the moving part. The coupling of the meshes between the moving parts and the stationary parts is handled by either the sliding surface method or the moving band method depending on the motion type. For the sliding surface method, the stationary and moving meshes are coupled together along the sliding interface of two coupling (master and slave) surfaces. For the moving band method, an additional band region is used to separate the stationary and moving parts and only the mesh in the band region is recreated at each time step. To achieve maximum flexibility and good discretization quality, non-conforming meshes are used for the coupling in both cases. This means that the scalar potential Ω at each node, the vector source field $\mathbf{H}_p$, and current vector potential $\mathbf{T}$ at each edge on the slave surface have to be mapped onto the master surface to eliminate all unknowns on the slave surface. Finding a general algorithm for mapping node-based scalar potential unknowns should not be too difficult. In the case of mapping vector unknowns, however, the process of splitting slave edge variables with respect to the trace of the master mesh while preserving the valid cutting domains is not impossible but very complicated.

To overcome this difficulty, a separation technique can be applied to confine the generation of every cutting domain to either the stationary region or the moving region without crossing the sliding interface [14]. The physical basis of this scheme assumes that each excitation coil and each eddy-current-induced conductor either entirely resides on the stationary part or entirely resides on the moving part but not on both. This means that the sum of the source or eddy currents enclosed by the sliding interface or the band is zero. This assumption is always true for practical problems. As a result, the process of splitting slave edges with respect to the trace of the master mesh for mapping edge-based vector potentials is completely eliminated and only the node-based scalar potential is involved. This separation scheme also has significant computational advantages for voltage-driven sources because it makes the non-zero support of each cutting domain much smaller and the equation system much sparser due to the support of current vector potential $\mathbf{T}$ being considerably reduced. In addition, the source field computation and cutting domain generation need to be done only once at the beginning of the entire transient simulation.

However, the generation of cutting domain is an automatic and random process, and the cutting edges may lay on or even spread across the sliding interface, which leads to the failure of cutting domain generation. To this end, we need to develop an algorithm to ensure that the generation of every cutting domain will reside on either the stationary region or on the moving region without crossing the sliding interface so that cutting edges will not lay on or spread across the sliding interface. The work reported in [15] proposed an algorithm that is able to guarantee that the generation of every cutting domain associated with both the source component $\mathbf{H}_S$ and the current vector potential **T** will reside on either the stationary region or on the moving region without crossing the sliding interface. This makes **T**--Ω formulation very efficient, reliable, and practical for the transient electromagnetic modeling with rigid motion.

Application Examples

The first example is a 3.7 kW, 8-pole three-phase induction motor. The motor has 48 stator slots and 44 rotor bars. Its Δ-connected three-phase winding is energized by 380 V at 50 Hz. Taking advantage of the periodic boundary condition, only two poles are modeled. The rotor cage with induced eddy current leads to a multiply connected domain. The developed algorithm has automatically and successfully confined the **T** cutting domain associated with cage to rotor and $\mathbf{H}_S$ cutting domain associated with three-phase windings to the stator even though both cage and three-phase windings are very close to the coupling surfaces. Figure 2.3 gives a flavor of what the automatically created cut domain looks like for one-phase winding. For the cage, 11 cuts associated with 11 holes are automatically identified even though one hole is cut into two halves in the axial direction. Figure 2.4 is the vector arrow plot of the induced eddy current at $t = 0.004$ s. Figure 2.5 shows B plot in stator, the induced eddy current in the bar and source current in the windings at $t = 0.0132$ s.

The second example is a 538 kW, 6-pole three-phase synchronous motor. This motor has 72 stator slots. Its Y-connected three-phase winding is energized by 400 line-to-line volts at 50 Hz. The damper (cage) will induce eddy current. As expected,

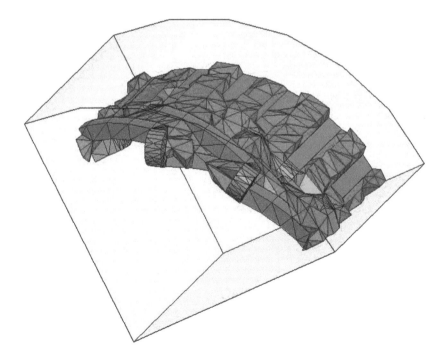

Figure 2.3   One-phase winding and its cutting domain.

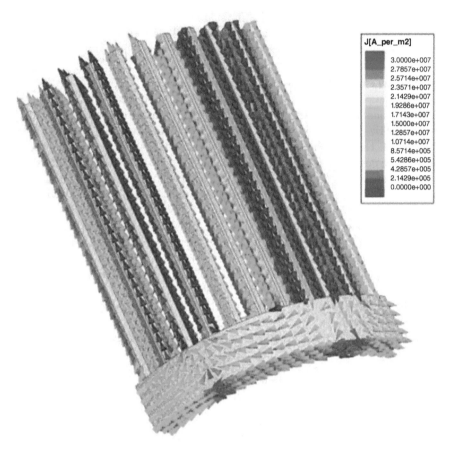

Figure 2.4   Induced eddy currents in the bar at $t = 0.004$ s.

both **T** cutting domain associated with the rotor damper and $\mathbf{H}_S$ cutting domain associated with three-phase windings on the stator and the excitation winding on the rotor are correctly created without crossing the coupling surfaces. For the cage, 7 cuts associated with 7 holes are identified automatically. Figure 2.6 shows the vector arrow plot of the induced eddy current at $t = 0.0088$ s. Figure 2.7 shows **B** plot in both stator and rotor, the induced eddy currents in the bar and source currents in the windings at $t = 0.0464$ s.

2.2   FIELD-CIRCUIT COUPLING

There exist two basic approaches to couple FEA with circuit simulation. One is the direct coupling approach [16–18] where the field and circuit equations are coupled directly together and solved simultaneously. The other approach is co-simulation by

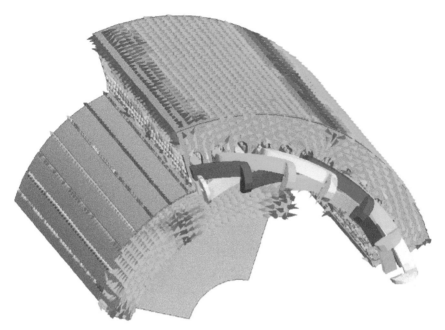

Figure 2.5   Flux density plot in the stator, induced eddy current plot in the bar, and the source currents in the windings at $t = 0.0132$ s.

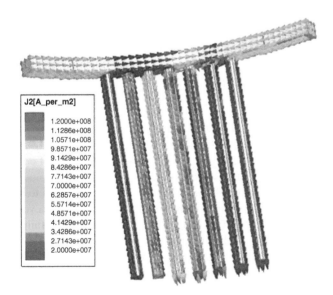

Figure 2.6   Induced eddy currents in the bar at $t = 0.0088$ s.

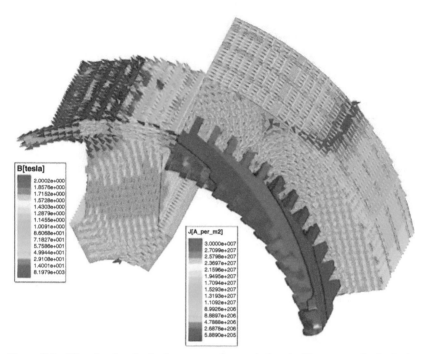

Figure 2.7   Flux density plot in the stator and rotor, induced eddy current plot in the bar and the source currents in the windings at $t = 0.0464$ s.

indirect coupling, where the FEA and circuit simulator are treated as separate systems in a time-stepping process, while they exchange coupling coefficients in each step [19, 20]. Indirect coupling becomes attractive when the time constants in the field domain and in the circuit domain differ significantly from each other. It also makes the development of individual simulators more efficient and easier to use by the experts in the different fields.

A common mechanism for indirect coupling is to use sources as coupling coefficients, where the field simulator uses the voltages across coupling windings as input and winding currents as output, while the circuit simulator uses the currents of coupling windings as input and winding voltages as output [21]. This approach provides the possibility of system level simulation. However, forcing coupling sources to be constant during a time step introduces potential convergence problems unless a very small time step is used. In addition, constant coupling current source may fail to model load commutation when a diode is directly connected to a winding. This may lead to an unphysical oscillation of the winding current or even to incorrectly keeping a diode in a permanent off state, if initial winding current is zero.

In this section, an indirect approach with parameter coupling is introduced [22]. In this approach, the FEA simulator provides the circuit simulator with lumped field parameters for every coupling winding, and the circuit simulator feeds back the Thevenin equivalent parameters to the FEA simulator, instead of forwarding voltages and currents between the two simulators. In this way, both FEA and circuit simulators can work independently. Since winding currents and voltages are both free to change across coupled windings in both the FEA and circuit simulators at each time step, the solution is more accurate and stable, and therefore, it may not require an iteration between the FEA and circuit simulators.

2.2.1 Circuit Parameter Extraction

Assume the discrete field equation with voltage-driven windings is finally expressed as

$$\begin{bmatrix} S_{11} & S_{1w} \\ S_{1w}^T & S_{ww} \end{bmatrix} \cdot \begin{bmatrix} X \\ I_w \end{bmatrix} = \begin{bmatrix} Y \\ V_w \end{bmatrix}, \tag{2.24}$$

where S is the stiffness matrix, X represents all edge unknowns associated with electric vector potential $\mathbf{T}$ and node unknowns associated with magnetic scalar potential Ω, I_w depicts the currents of all voltage-driven windings w, V_w denotes all voltage-driven excitations, and Y relates to the contribution of all other excitations. When all voltage-driven windings are excited independently, V_w is known, and equation (2.24) can be solved directly.

For multiphase, such as three-phase, Y-connected windings, even though the voltage source can be expressed independently for each winding, since the common node of windings is not connected with that of the voltage sources, the voltage difference between the two common nodes may not be zero, therefore, V_w in equation (2.24) is unknown. In such a case, a branch-to-loop transformation can be used to transfer the branch currents I_w to loop currents I_l, and equation (2.24) becomes

$$\begin{bmatrix} S_{11} & S_{1w}B_f^T \\ B_f S_{1w}^T & S_{ll} \end{bmatrix} \cdot \begin{bmatrix} X \\ I_l \end{bmatrix} = \begin{bmatrix} Y \\ V_{ls} \end{bmatrix}, \tag{2.25}$$

where B_f is the branch-to-loop transformation matrix, and

$$\begin{cases} S_{ll} = B_f S_{ww} B_f^T \\ V_{ls} = B_f V_w \end{cases} \tag{2.26}$$

After equation (2.25) is solved, the branch (winding) currents can be obtained from

$$I_w = B_f^T I_l. \tag{2.27}$$

Equation (2.25) can be regarded as a general format because equation (2.24) is actually a special case of equation (2.25) when B_f is an identity matrix.

For the winding set w, if voltage sources are not directly applied across winding terminals, instead, they are indirectly applied to windings via a complicatedly connected circuit, the winding terminal voltages are unknown. In such a case, V_{ls} and S_{ll}, together with the branch-to-loop transformation matrix B_f, must be provided by the circuit simulator.

For any component in the circuit, the discrete format of branch voltage equation can be generalized as

$$r_b(i_b + i_{bs}) = v_b + v_{bs},$$ (2.28)

where r_b is the branch equivalent resistance, i_{bs} and v_{bs} are the branch equivalent current and voltage sources, respectively, and i_b is the branch current and v_b is the branch terminal voltage. The branch equivalent sources may be obtained directly from the excitation, and/or derived from the initial condition, and/or from the intercept of a load line tangent to the nonlinear characteristic.

All branch voltage equations in the circuit can be expressed in the matrix form, and coupled with the field equation (2.24) as

$$\begin{bmatrix} S_{11} & S_{1w} & 0 \\ S_{1w}^T & S_{ww} & 0 \\ 0 & 0 & S_{bb} \end{bmatrix} \cdot \begin{bmatrix} X \\ I_w \\ I_b + I_{bs} \end{bmatrix} = \begin{bmatrix} Y \\ V_w \\ V_b + V_{bs} \end{bmatrix}$$ (2.29)

If the winding branch set w and the circuit branch set b are combined as total branch set B, equation (2.29) can be rewritten as

$$\begin{bmatrix} S_{11} & S_{1B} \\ S_{1B}^T & S_{BB} \end{bmatrix} \cdot \begin{bmatrix} X \\ I_B + I_{Bs} \end{bmatrix} = \begin{bmatrix} Y \\ V_B + I_{Bs} \end{bmatrix}$$ (2.30)

In equation (2.30), V_B is unknown, and it must be eliminated by applying Kirchhoff's current and voltage laws. There are two algorithms in circuit analysis, one is called nodal analysis and the other is called loop analysis.

Nodal Analysis
By introducing branch-to-node transformation matrix A, the branch voltages V_B can be obtained from the nodal voltages V_N as

$$V_B = A^T V_N$$ (2.31)

and the Kirchhoff's Current Law (KCL) is expressed as

$$A I_B = 0$$ (2.32)

Combining equations (2.30) –(2.32) to eliminate V_B, we obtain

$$\begin{bmatrix} S_{11} + S_{11}' & S_{1B} S_{BB}^{-1} A^T \\ A S_{BB}^{-1} S_{1B}^T & A S_{BB}^{-1} A^T \end{bmatrix} \cdot \begin{bmatrix} X \\ V_N \end{bmatrix} = \begin{bmatrix} Y \\ I_{Ns} \end{bmatrix},$$ (2.33)

where

$$\begin{cases} S_{11}' = S_{1B} S_{BB}^{-1} S_{1B}^T = S_{1w} S_{ww}^{-1} S_{1w}^T \\ I_{Ns} = A(S_{BB}^{-1} V_{Bs} - I_{Bs}) \end{cases}$$ (2.34)

Nodal analysis obtains nodal voltages of the circuit based on equation (2.33), from which branch voltages are obtained by equation (2.31). To solve equation (2.33), we need to modify field stiffness matrix S_{11} (adding S'_{11} to S_{11}), which is very costly due to its large size.

Loop Analysis

Introducing branch-to-loop transformation matrix B, we get the branch currents I_B from the loop currents I_L as

$$I_B = B^T I_L \tag{2.35}$$

The Kirchhoff's Voltage Law (KVL) is expressed as

$$B V_B = 0 \tag{2.36}$$

Eliminating V_B in equation (2.30) based on equations (2.35)–(2.36), we get

$$\begin{bmatrix} S_{11} & S_{1B}B^T \\ BS_{1B}^T & S_{LL} \end{bmatrix} \cdot \begin{bmatrix} X \\ I_L \end{bmatrix} = \begin{bmatrix} Y \\ V_{Ls} \end{bmatrix}, \tag{2.37}$$

where

$$\begin{cases} S_{LL} = B S_{BB} B^T \\ V_{Ls} = B(V_{Bs} - S_{BB} I_{Bs}) \end{cases} \tag{2.38}$$

Loop analysis solves loop currents of the circuit, from which branch current can be obtained by equation (2.35). Comparing equations (2.37) and (2.33), we find that using the loop analysis is much more convenient than using the nodal analysis because we do not need to modify field stiffness matrix S_{11}, and the expressions for other matrix components are simple. Therefore, loop analysis is employed in our coupling approach.

Thevenin Equivalent Parameters

In equation (2.37), the size of S_{LL} is $L \times L$, where L is the number of total loops. This size may be very large, depending on the circuit. Using Thevenin equivalent parameters, this size can be significantly reduced.

From equation (2.37), we can write the circuit equation as

$$B S_{1B}^T X + S_{LL} I_L = V_{Ls} \tag{2.39}$$

As has been described above, the total branch set B consists of winding branch subset w and circuit branch subset b. Comparing equations (2.29) and (2.30), we have

$$S_{1B} = \begin{bmatrix} S_{1w} & 0 \end{bmatrix} \tag{2.40}$$

If we decompose the total loop set L into two loop subsets, that is winding loop subset l and circuit loop subset c, we have

$$B = \begin{bmatrix} B_{lw} & B_{lb} \\ B_{cw} & B_{cb} \end{bmatrix} \tag{2.41}$$

If we define all circuit loops in such a rule that: a winding loop current may go through some circuit branches, but all circuit loop currents will not go through a winding branch, then we have $B_{cw} = 0$.

Now, equation (2.39) can be rewritten as

$$\begin{bmatrix} B_{lw} & B_{lb} \\ 0 & B_{cb} \end{bmatrix} \begin{bmatrix} S_{1w}^T \\ 0 \end{bmatrix} X + \begin{bmatrix} S_{ll}' & S_{lc} \\ S_{lc}^T & S_{cc} \end{bmatrix} \begin{bmatrix} I_l \\ I_c \end{bmatrix} = \begin{bmatrix} V_{ls}' \\ V_{cs} \end{bmatrix} \tag{2.42}$$

Eliminating I_c in equation (2.42), we get

$$B_{lw} S_{1w}^T X + (S_{ll}' - S_{lc} S_{cc}^{-1} S_{lc}^T) I_l = V_{ls}' - S_{lc} S_{cc}^{-1} V_{cs} \tag{2.43}$$

Considering B_{lw} is actually the same as B_f in equation (2.25), and combining equation (2.43) with the FEA equation in (2.37), we obtain

$$\begin{bmatrix} S_{11} & S_{1w} B_f^T \\ B_f S_{1w}^T & S_{ll} \end{bmatrix} \cdot \begin{bmatrix} X \\ I_l \end{bmatrix} = \begin{bmatrix} Y \\ V_{ls} \end{bmatrix} \tag{2.44}$$

where

$$\begin{cases} S_{ll} = S_{ll}' - S_{lc} S_{cc}^{-1} S_{lc}^T \\ V_{ls} = V_{ls}' - S_{lc} S_{cc}^{-1} V_{cs} \end{cases} \tag{2.45}$$

Equation (2.44) has the same form as equation (2.25), but parameters are given by equation (2.45), which are actually the Thevenin equivalent parameters.

2.2.2 Field Parameter Extraction

On the finite element side, let the winding flux linkage λ be split into two components. One is Li, produced by coupling branch currents through the winding inductances. The other is *internal* flux linkage, ψ, produced by other sources, such as permanent magnets, other windings not connected to the coupling nodes, and induced eddy currents. Thus, the induced *terminal* voltage can be expressed as

$$\begin{aligned} e_t &= \frac{d\lambda}{dt} = \frac{d(Li)}{dt} + \frac{d\psi}{dt} \\ &= L_1 \frac{\Delta i}{\Delta t} + i_0 \frac{\Delta L}{\Delta t} + \frac{d\psi}{dt} \end{aligned} \tag{2.46}$$

This can also be written as

$$\begin{aligned} e_t &= L_1 \frac{di}{dt} + \frac{(L_1 i_0 + \psi_1) - (L_0 i_0 + \psi_0)}{\Delta t} \\ &= L_1 \frac{di}{dt} + \frac{\lambda_1' - \lambda_0}{\Delta t} = L_1 \frac{di}{dt} + e_i \end{aligned} \tag{2.47}$$

where

$$e_i = \frac{\lambda_1' - \lambda_0}{\Delta t} \tag{2.48}$$

with $\Delta t = t_1 - t_0$.

Subscripts "1" and "0" stand for the current time point and the previous time point, respectively, and e_i is the induced *internal* voltage due to permanent magnet, motion, eddy current, the excitation of other windings not connected to the coupling circuit and the contribution from dL/dt. λ_1' in equation (2.48) is derived under the same conditions as λ_1, except replacing the excitation of each coupling winding by current source with the values at previous time point. Note that the eddy effects due to the change in the winding currents have been accounted for in the equivalent inductance. The eddy effects due to other factors, such as motion, magnet, and other internal sources are accounted for in the induced *internal* voltage e_i. Thus, the Thevenin equivalent of windings can be represented by an inductance L in series with an internal voltage e_i. In addition, R is used to represent either the stranded winding resistance or the solid-conductor winding resistance which is derived from the field solutions.

After FEA at each time step, the FEA system's coefficient matrix is frozen, which is equal to freezing the permeability of each element. In this way, the flux density distribution corresponding to each individual excitation can be solved separately based on the superposition principle for linear field problems. In 3D analysis using a **T–Ω** formulation, the flux linkage in winding k caused by the distributed flux density **B** is derived from [18]

$$\lambda_k = \iiint_{R_k} \mathbf{H}_k \cdot \mathbf{B} \, dR, \tag{2.49}$$

where $\mathbf{H}_k$ is the field corresponding to 1 A current in winding k. Note that $\mathbf{H}_k$ is just a distributed vector coefficient, it does not really contribute to **B** distribution.

In equation (2.49), if **B** is created by an individual excitation of 1 A current in winding j (individual excitation means all other excitations including PMs are zero), the computed flux linkage in winding k is actually the mutual inductance between windings j and k. In this way, all self and mutual inductances in the inductance matrix can be computed. If **B** is created by such excitations that all windings connected to the coupling circuit are injected with the currents at the previous time point and all other excitations keep the values of the current time point, the computed flux linkage is λ_1' based on which the *internal* voltage of winding k can be derived from equation (2.48). Then, if currents of all windings connected to the coupling circuit are replaced by the values at the current time point, the computed flux linkage is λ_1 which can be stored for *internal* voltage computation in next time step.

After field parameters have been extracted in FEA, field simulator will pass all these parameters, together with solved currents of the current time point and other necessary information such as time step and rotor position, to circuit simulator. The circuit simulator will inject all winding currents into the circuit, which has been frozen based on the previous circuit solution, to update all branch voltages/currents and output all specified voltage/current data at the current time point. Then, the circuit simulator will predict the circuit solution at the next time point (assuming all field parameters keep constant in this time step), freeze the circuit, and feed the predicted Thevenin equivalent parameters for the next time point back to the field simulator. This process will continue step by step until the stop time is reached.

2.2.3 Adaptive Time Step

In order not to increase the computation time caused by field-circuit coupling, the circuit simulator will use the same time step as FEA, ΔT in default. However, if some cases happen which may cause sudden changes in current and/or voltage waveforms, such as ON/OFF status change of a switch or diode, during this time step, the circuit simulator will reduce the time step to Δt to avoid such changes happen within the time step. This reduced time step will be fed back to the field simulator, and the transient field is solved with the reduced time step. In this way, the circuit ON/OFF status changes will be caught even though the originally specified time step in FEA ΔT is quite large.

Assume the specified minimum circuit time step is Δt_{min}, following rules are applied to adaptively change the time step:

a. If the previous time step is reduced based on the ON/OFF time interface, let $\Delta t = \Delta t_{min}$, change the ON/OFF status for the related component, and skip the following steps;

b. For each voltage or current source, if the waveform is PULSE (pulse wave) or PWL (piecewise linear), each corner point is a must-run time point. If there exist any must-run time points in all voltage and current sources, reduce the time step based on the earliest must-run time point;

c. In this reduced time step Δt, if the ON/OFF status of the end time point is different with that of the start time point for all switches and diodes, find the earliest ON/OFF time interface using the binary search, update the reduce time step Δt, and keep status unchanged for all components in this time step;

d. If $\Delta t < \Delta t_{min}$, let $\Delta t = \Delta t_{min}$.

As an application example, a 4-pole three-phase spindle motor is simulated. As shown in Figure 2.8, the stator is fed by a DC–AC inverter and the rotor is excited by permanent magnets. The chopped-current control through controlled switches T1–T6 maintains stator currents within the hysteresis band 8 ± 1 A as shown in Figure 2.9. This example demonstrates the effectiveness of the adaptive time-stepping algorithm, which allows the default FE time step to be set to a very large 0.001 s. However, once switching is detected, the FE time step is automatically reduced. The dots on the curves represent every computation instant with reduced minimum time step 2×10^{-5} s.

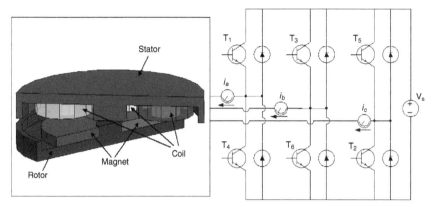

Figure 2.8   Spindle motor fed by inverter with chopped-current control.

2.2.4   Brush-Commutation Model for DC Machines

With the field-circuit coupling approach, it is quite convenient to develop some dedicated components for electric machine applications. The circuit brush-commutation model can be one good example of such dedicated components.

Commutation phenomena present a challenging problem in modeling the performance of brush-commutation machines such as DC machines and universal motors. Some field-circuit coupling approaches were developed in [23–26] to simulate commutation behavior. Reference [23] models commutation using an alternative polarity index as a function of position, and couples the commutation equations with transient FEA. In this approach, designers must define polarity indexes as

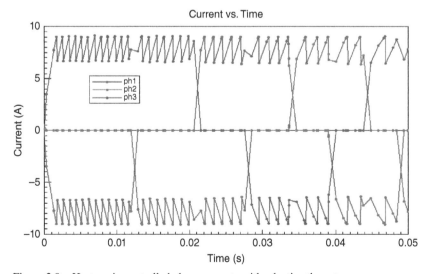

Figure 2.9   Hysteresis-controlled phase currents with adaptive time step.

a function of position for all armature coils. This task is cumbersome and needs special treatment. In [25], the armature winding is separated into two parts, one main winding and one or more commutating windings, based on the commutation status. The inductance matrix of the main winding and the commutating windings is computed using static field analysis at different saturation levels and different rotor positions for one commutating cycle. The stored inductance matrix together with the commutation equations is then analyzed in the circuit domain. Since the size of the inductance matrix and the commutation equations depend on the number of commutator segments covered by the brushes, this approach is difficult to generalize.

In Subsection 2.2.4, a circuit component called "commutator bar" [27] is introduced to model the position relationship between each commutator segment and each brush. All commutator bar components are associated with a commutation model that describes the commutation characteristic. With the help of these new components, the commutating circuit can be created by simply connecting all coils to the corresponding commutator bar components and selecting the commutation model. The required complicated commutation equations can be automatically created based on equations (2.44) –(2.45) from this commutation circuit without tedious work.

Commutation Circuit

For clarity, a 2-pole 12-slot permanent magnet DC (PMDC) motor with a lap armature winding is used to describe the procedure. The armature winding has a 5-slot coil pitch, as shown in Figure 2.10. The flat expanded view of the motor is shown in Figure 2.11 indicating the spatial relationship between the permanent magnets, coils, commutator segments, and brushes at the initial position. With the indicated direction of rotation, the brush aligned with the south pole is positive and that aligned with the north pole is negative. The "go" terminal of *coil0* connects to *seg0* and its return terminal connects to *seg1*; the "go" terminal of *coil1* connects to *seg1* and its return terminal connects to *seg2*, etc.

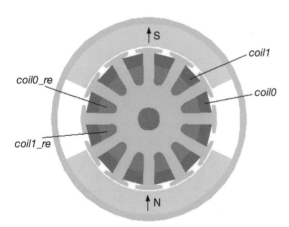

Figure 2.10   The geometry layout of the sample PMDC motor.

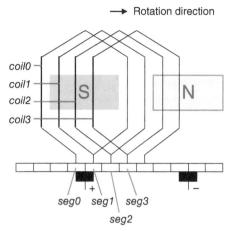

Figure 2.11   The flat expanded view of the sample PMDC motor.

At the initial position shown in Figure 2.11, *seg0* lags the positive brush by 15 mechanical degrees (one half of a commutator segment pitch), and it lags the negative brush by 195 mechanical degrees; *seg1* lags the positive brush by −15 mechanical degrees or 345 degrees, and it lags the negative brush by 165 mechanical degrees, etc. In Figure 2.12, *Bar0_pos* labels the connection of *seg0* to the positive brush and *Bar0_neg* labels that to the negative brush, etc. The lagging angle of the initial position is a parameter of the commutator bar component. Another parameter of the commutator bar component is the instance name of the associated commutation model, as shown in Table 2.1.

Commutation Model

In Figure 2.12, an instance of the commutation model named *ComModel* is introduced to define the commutation characteristic. For simplicity but without losing generality, a linear model is used to model the brush voltage drop as a linear function of the contact current. As shown in Figure 2.13, the contact conductance between a brush and a commutator segment is proportional to the contact area, and is a periodic function of rotor position. The current rotor position is obtained from the transient FEA solver.

In Figure 2.13, the initial lagging angle parameter *LagAngle* is obtained from the commutator bar component, and its value is different for each individual component. All other parameters (*Gmax*, *WidB*, *WidC*, and *Period*) are specified in the commutation model and their descriptions are given in Table 2.2. Positions a, b, c, and d in Figure 2.13 correspond to four positions when one extremity of a commutator segment aligns with one extremity of a brush, as shown in (a), (b), (c) and (d) of Figures 2.14 and 2.15, respectively.

Simulation Results

A PMDC motor with the geometry as shown in Figure 2.10 is simulated at load operating conditions using the above described field-circuit coupling method. The

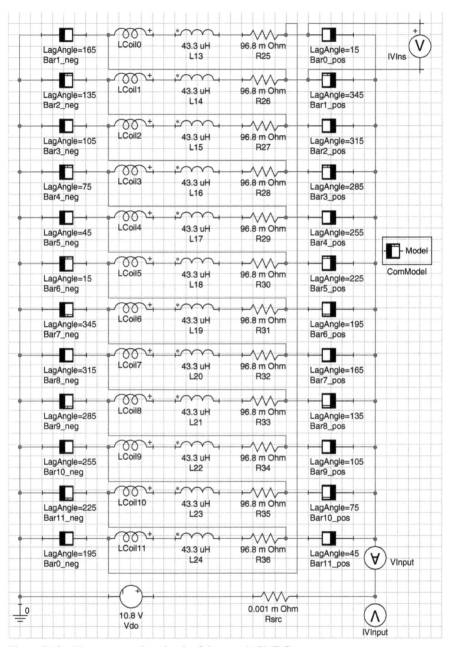

Figure 2.12   The commutating circuit of the sample PMDC motor.

TABLE 2.1   Parameters of commutator bar component

Parameters	Descriptions
MOD	The associated commutation model instance name
LagAngle	Lagging angle, in mechanical degrees, of the specified commutator segment to the specified brush at initial time

simulated results are compared with the measured data in Table 2.3. Some important information associated with commutation process can be obtained from simulation. The simulated current of *coil0*, compared with that of brush shift of 10 mechanical degrees in the anti-rotating direction, is shown in Figure 2.16. It can be observed, as the common understanding, that the commutation with 10° brush shift is improved because the current commutates more linearly than that without brush shift. The terminal voltage of *coil0*, or the voltage over the insulation between *seg0* and *seg1*, is displayed in Figure 2.17. It is obvious that the maximum voltage of a coil, an

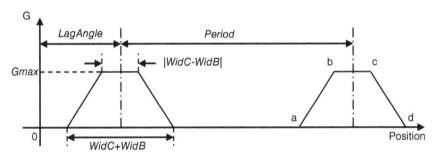

Figure 2.13   Contact conductance as a periodic function of rotor position.

TABLE 2.2   Parameters of commutation model

Parameters	Descriptions
Gmax	Full brush-commutator contact conductance
WidB	Brush width in mechanical degrees
WidC	Commutator segment width in mechanical degrees
Period	Period of a commutator segment rotating from one brush to another brush of the same polarity in mechanical degrees

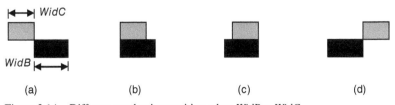

| (a)   (b)   (c)   (d)

Figure 2.14   Different conducting position when *WidB* > *WidC*.

(a) (b) (c) (d)

Figure 2.15   Different conducting position when *WidB* < *WidC*.

TABLE 2.3   Parameters of commutation model

	Simulated results	Measured data
Speed	3162 rpm	3162 rpm
Torque	0.274 Nm	0.27 Nm
Current	15.1 A	14.94 A
Peak–peak current	2.06 A	2.02 A

important quantity used to determine the thickness of segment insulation, appears just after that coil has finished commutating.

2.3   FAST AC STEADY-STATE ALGORITHM

In many cases, what electrical machine designers are interested in is the steady-state performance, instead of the transient process, at AC voltage excitations. The transient

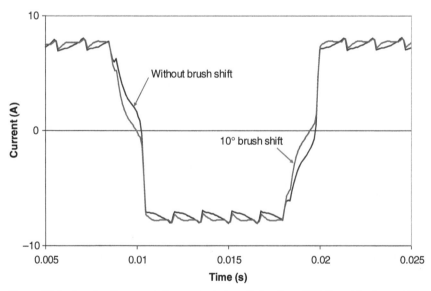

Figure 2.16   Load coil commutating current at applied voltage 10.8 V and load speed 3162 rpm compared with that of 10° brush shift.

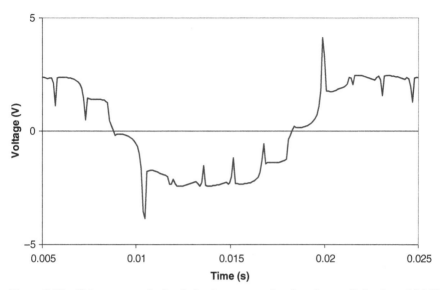

Figure 2.17   Voltage across the insulation between *seg0* and *seg1* at applied voltage 10.8 V and load speed 3162 rpm.

simulation of an electrical machine often requires simulating much more than one period in order to obtain the periodic steady-state behavior, depending on the time constant of the electromagnetic system. A lot of published methods have been developed to get the steady-state solution in the literature fast, which are mainly categorized into two methodologies, one obtains the steady-state solution directly and the other speeds up the transient process by proper initial conditions.

One widely used method to directly obtain the periodic steady-state solution is the harmonic balance method using an eddy current solver in the frequency domain [28]. The unknown potentials are represented by Fourier-series and the nonlinear behavior of the material is split into a linear and a nonlinear term using a fixed-point technique. The nonlinear term will cause harmonic excitations. All fundamental and harmonic excitations are updated based on the previous solution in an iteration process until the resultant field properties satisfy the nonlinear behavior. This method is computationally efficient because each linear matrix equation related to each individual harmonic excitation can be solved independently by parallel computation. However, it is difficult to employ this method in electrical machine analysis because different frequencies may exist in both the static and moving parts for an individual frequency excitation.

Another method to directly obtain the steady-state solution is to solve all time steps of one period together with the periodic condition in the time domain [29]. Using time decomposition method (TDM), the reverse of linearized stiffness matrix of each time step can be independently solved by parallel computation, and then solutions of all time steps are obtained. A nonlinear iteration process is required to update equations based on the previous solutions of all time steps until convergence is reached. The details are illustrated in Section 2.4.

One possible technique to speed up the transient process is using the frequency-domain solution to estimate the initial values [30]. This method is efficient only when the following both conditions are satisfied: (1) under sinusoidal voltage excitation in a nonlinear system, all harmonic current components are negligible, such as in three-phase induction motors; (2) with the problem involving eddy currents, not only all winding currents, but also eddy current density distribution, must be passed from the AC FEA solver to the transient FEA solver as initial values.

To speed up the transient process, another method called "phase balancing" method [31] was developed to repeatedly modify the initial currents in each successive period by partially eliminating the DC currents based on the previous-period solution. This iteration process continues until the DC current components are totally eliminated in all phases. The DC currents of the previous period are computed under the assumption that the AC current waveforms with DC bias are the same as those without DC bias. Due to the nonlinearity, the DC currents may be overestimated. Therefore, an under-relaxation factor was used to ensure convergence, which causes the process to require several periods to eliminate the DC currents. Not applicable to the problems with eddy currents is another disadvantage of this method.

In this section, an efficient method, referred to as alternating flux linkage (AF) model, will be introduced to speed up the transient process for reaching fast AC steady state, almost suitable for all AC machines. The DC flux linkage is eliminated by applying an additional voltage component within a small time interval [32], such as in a half period, for each phase. The flux linkages at the end of this interval are the perfect initial values for flux linkages in the succeeding time. As a result, the steady-state AC flux linkages will be obtained immediately within the next period.

2.3.1 Alternating Flux Linkage Model

Basic Equations

Assuming a winding is applied with an AC voltage source of

$$v(t) = V_m \sin(2\pi f t + \varphi), \tag{2.50}$$

if the winding resistance is negligible, the flux linkage of the winding can be computed from the applied voltage as

$$\lambda(t) = \lambda(0) + \int_0^t V_m \sin(2\pi f t + \varphi) dt$$
$$= [\lambda(0) + \lambda_m \cos(\varphi)] - \lambda_m \cos(2\pi f t + \varphi), \tag{2.51}$$

where $\lambda(0) + \lambda_m \cos(\varphi)$ is the DC component with $\lambda(0)$ being the initial flux linkage, and λ_m, the amplitude of AC component, is derived from

$$\lambda_m = V_m/(2\pi f) = V_m T/(2\pi) \tag{2.52}$$

with $T = 1/f$ being the period. In transient FEA solvers, the initial field solution can be obtained via *static field analysis* based on the initial winding currents. Usually in permanent magnet (PM), or field excited, machines, the initial winding flux linkage is not zero, and its effects must be taken into account.

After the DC component of the flux linkage decays to zero, the flux linkage will reach the steady-state value. If the applied voltage is modified in such a way that the DC component of the flux linkage is zero after time t_s, the steady state can be immediately reached for $t > t_s$.

Assume an additional voltage component of

$$v_d(t) = \begin{cases} \delta(t) & t < t_s \\ 0 & t \geq t_s \end{cases} \tag{2.53}$$

is applied to the winding so that the DC component of the flux linkage at t_s ($t_s>0$) is zero. In such a case, the flux linkage for $t > t_s$ is

$$\lambda(t) = \lambda(0) + \int_0^t [v(t) + v_d(t)]dt$$

$$= -\lambda_m \cos(2\pi f t + \varphi) + \lambda(0) + \lambda_m \cos(\varphi) + \int_0^{t_s} \delta(t)dt \tag{2.54}$$

In equation (2.54), $\delta(t)$ can be any function and t_s can be freely selected. As long as

$$\int_0^{t_s} \delta(t)dt = -[\lambda(0) + \lambda_m \cos(\varphi)] \tag{2.55}$$

the DC component of the flux linkage at t_s will be zero, and the flux linkage will reach the steady state after time t_s. One possible select is

$$\begin{cases} t_s = T/2 \\ \delta(t) = V_{dm} \sin(2\pi f t) \end{cases} \tag{2.56}$$

To satisfy equation (2.56), we have

$$\frac{2}{2\pi f} V_{dm} = -[\lambda(0) + \lambda_m \cos(\varphi)] \tag{2.57}$$

or

$$V_{dm} = -\frac{\pi}{T} \lambda(0) - \frac{1}{2} V_m \cos(\varphi) \tag{2.58}$$

For poly-phase windings, the applied AC voltages can be modified phase by phase independently.

This algorithm is based on the assumption of negligible winding resistance. If the winding resistance is not negligible, the DC flux linkage at t_s will not be zero.

The larger the resistance is, the larger error the DC flux linkage will have at t_s, but fortunately, the faster the error DC flux linkage will decay to zero due to small time constant. Therefore, when winding resistance is large, it may need a couple of additional periods to reach to the steady state.

Effects of Time Discrete

The above equations, which are derived from the integration of continuous functions, will be accurate enough for discrete computation if time step is small. If the time step is too large so that its impact is not negligible, the applied additional voltage needs to be adjusted to consider the time-step effects. When backward Euler method is employed in FEA, the discrete format of the induced voltage is

$$v(t_k) \approx e(t_k) = \frac{\lambda(t_k) - \lambda(t_{k-1})}{t_k - t_{k-1}} \tag{2.59}$$

If $t_k - t_{k-1} = \Delta t$ is constant, the flux linkage will be

$$\lambda(t_k) \approx \lambda(t_{k-1}) + v(t_k) \cdot \Delta t$$

$$= \left[\lambda(0) + \sum_{i=1}^{k} \frac{v(t_{i-1}) + v(t_i)}{2} \Delta t - \frac{v(t_k) + v(0)}{2} \Delta t \right] + v(t_k) \cdot \Delta t$$

$$\approx \lambda(0) + \int_0^{t_k} v(t) dt + \frac{v(t_k) - v(0)}{2} \Delta t$$

$$= -\lambda_m \cos(2\pi f t_k + \varphi)$$
$$+ \lambda(0) + \lambda_m \cos(\varphi) + \frac{1}{2} \left[v(t_k) - v(0) \right] \Delta t \tag{2.60}$$

Let t_k to be the time point at which $v(t_k) = 0$, that is

$$2\pi f t_k + \varphi = 2\pi \tag{2.61}$$

then,

$$\lambda_m \cos(2\pi f t_k + \varphi) = \lambda_m \tag{2.62}$$

Hence, equation (2.60) becomes

$$\lambda(t_k) = -\lambda_m + \lambda(0) + \lambda_m \cos(\varphi) - \frac{1}{2} v(0) \Delta t \tag{2.63}$$

Therefore, as long as the additional applied voltage $v_d(t)$ satisfies

$$\int_0^{t_s} \delta(t) dt = -[\lambda(0) + \lambda_m \cos(\varphi) - \frac{1}{2} v(0) \Delta t] \tag{2.64}$$

the flux linkage will reach its extreme, or negative peak, value $-\lambda_m$ at time t_k, that is

$$\lambda(t_k) = -\lambda_m \tag{2.65}$$

On the other hand, the peak-to-peak flux linkage after time t_k can be computed from

$$\lambda_{p2p} = \int_{t_k}^{t_k+T/2} v(t)dt = 2\lambda_m \tag{2.66}$$

The above two equations show that the flux linkage will vary between $-\lambda_m$ and $+\lambda_m$, or reach the steady state, after time t_k when equation (2.64) is satisfied. From equations (2.57) and (2.64), we get

$$V_{dm} = -\frac{\pi}{T}\lambda(0) - \frac{1}{2}V_m\cos(\varphi) + \frac{\pi \cdot \Delta t}{2T}V_m\sin(\varphi) \tag{2.67}$$

Based on equation (2.67), the additional voltage component as expressed by

$$v_d(t) = \begin{cases} V_{dm}\sin(2\pi ft) & t < T/2 \\ 0 & t \geq T/2 \end{cases} \tag{2.68}$$

can be manually added to the AC voltage excitation of equation (2.50) for each phase in a transient FEA solver.

2.3.2 Applications in Direct AC Voltage Excitation

Algorithm for Arbitrary AC Voltage Sources
Equation (2.67) is derived based on the input voltage of sinusoidal waveform (equation 2.50), where the parameters V_m, f, and φ are directly obtained from the specification.

In real applications, a designer may specify an AC voltage for each winding with arbitrary expression. In such cases, the AC voltage can be evaluated at a series of discrete time points and expressed in a uniformly distributed table $v_i = f_{AC}(i\Delta t)$, where $f_{AC}()$ is the input expression of the AC voltage with $i = 0, 1, \ldots, n$, and $\Delta t = T/n$. The AC voltage expression can be validated by

$$\begin{cases} v_0 = v_n \\ \displaystyle\sum_{i=0}^{n-1} v_i = 0 \end{cases} \tag{2.69}$$

With backward Euler method, the flux linkages in the first period are obtained from

$$\lambda_k = \lambda_0 + \sum_{i=1}^{k} v_i \cdot \Delta t \quad k = 1, 2, \ldots, n-1, \tag{2.70}$$

where λ_0 is the initial flux linkage. The DC component of the flux linkage can be computed from

$$
\begin{aligned}
\lambda_{DC} &= \frac{1}{n}\left(\lambda_0 + \sum_{k=1}^{n-1}\lambda_k\right) \\
&= \frac{1}{n}\left(\lambda_0 + (n-1)\lambda_0 + \sum_{k=1}^{n-1}\sum_{i=1}^{k} v_i \cdot \Delta t\right) \\
&= \lambda_0 + \frac{T}{n^2}\left(\sum_{k=1}^{n-1}\sum_{i=1}^{k} v_i\right)
\end{aligned}
\tag{2.71}
$$

In order to eliminate the DC component of the flux linkage, the additional applied voltage must satisfy

$$
\int_0^{t_s} \delta(t)dt = -\lambda_{DC}
\tag{2.72}
$$

or

$$
V_{dm} = -\lambda_{DC} \cdot \pi/T
\tag{2.73}
$$

In equation (2.73), the effects of time discrete have been considered in equation (2.71).

Application Example

As an application example, a three-limb three-phase transformer is simulated at no-load operation. The 3D FEA model with 1/4 symmetry is shown in Figure 2.18. Without using the AF model, the applied three-phase voltage waveforms are shown in Figure 2.19, and the simulated three-phase flux linkage and current waveforms are shown in Figures 2.20 and 2.21, respectively.

Figure 2.21 shows that the inrush current at the time point of a half period (8.33 ms) reaches about 13 kA, because the flux linkage peak is doubled at this time point

Figure 2.18   The 3D FEA model of three-limb three-phase power transformer with 1/4 symmetry.

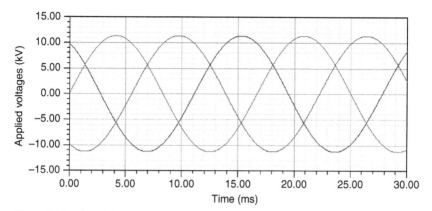

Figure 2.19   Applied three-phase voltage waveforms without using AF model.

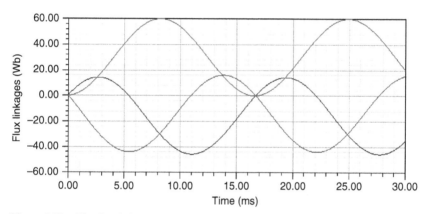

Figure 2.20   Simulated three-phase flux linkage waveforms without using AF model.

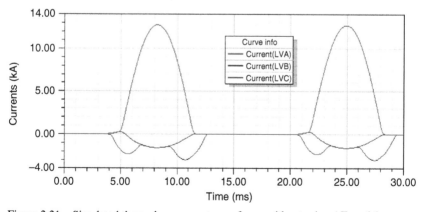

Figure 2.21   Simulated three-phase current waveforms without using AF model.

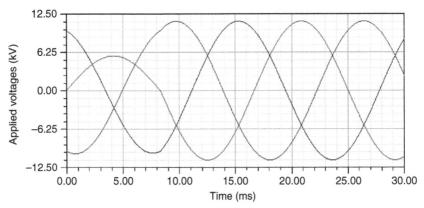

Figure 2.22   Applied three-phase voltage waveforms using the AF model.

due to the DC bias. It will take very long time for the DC flux linkage to decay to zero due to its large time constant.

To eliminate the DC flux linkage quickly, the applied three-phase voltages are modified from the input three-phase AC voltages based on the AF model illustrated above, as shown in Figure 2.22. The simulated three-phase flux linkage and current waveforms are shown in Figures 2.23 and 2.24, respectively. Figure 2.24 shows that using the AF model, the steady state is quickly reached in the second period.

2.3.3   Applications in Field-Circuit Coupling

Algorithm for Field-Circuit Coupling

In field-circuit coupling applications, the AF model can be specified in the circuit components of AC voltage sources. For example, a sinusoidal voltage source with the AF model can be specified as shown in Table 2.4. The parameters of the sinusoidal voltage source expressed in equation (2.50) can be directly obtained from the specification, and therefore, the additional voltage components for all three phases

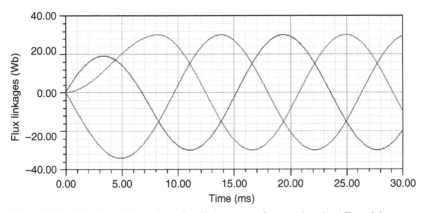

Figure 2.23   Simulated three-phase flux linkage waveforms using the AF model.

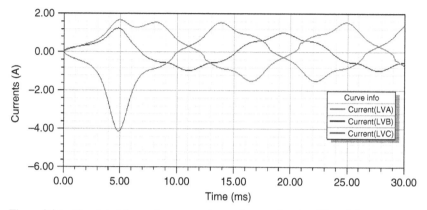

Figure 2.24 Simulated three-phase current waveforms using the AF model.

can be computed by equations (2.67)–(2.68). However, the initial flux linkage is related to each phase winding, and a voltage source may not be directly applied at the winding terminal, $\lambda(0)$ in equation (2.67) cannot be directly obtained from a winding.

In equation (2.41), for the winding loop subset l of all branches, if we pick up only the m branches related to AC voltage sources with AF model, we can construct a new branch-to-loop transformation matrix B_{lm} with the size of $l \times m$. If the flux linkages of all windings are expressed as a vector Λ_w, and flux linkages to be determined in all AC voltage sources with AF model are denoted as X_m, then we have

$$B_{lm}X_m = Y_l, \qquad (2.74)$$

where

$$Y_l = B_{lw}\Lambda_w \qquad (2.75)$$

The matrix B_{lm} is normally not a square matrix. Even though it is a square matrix in some special cases, it may not be full rank. Therefore, in general cases, there are infinite solutions of X_m. The simplest solution is obtained by setting all dependent variables be zero via following process:

a. Set all elements of X_m be zero;

b. During the Gaussian elimination, if a diagonal element $b_{ii} = 0$, try to find a non-zero element b_{jk} for $j \geq i$, $k \geq i$;

TABLE 2.4 **Specification of a sinusoidal voltage source with the AF model**

Parameter	Value	Description
Vm	11267.7	Peak amplitude, in Volts
Freq	50	Frequency, in Hz
Phase	0	Phase delay, in electrical degrees
AF	1	AF model, 1 for using AF mode

c. If $b_{jk} \neq 0$ is found, swap the rows of i and j for both B_{lm} and Y_l, swap columns of i and k for both B_{lm} and X_m, and continue the Gaussian elimination;

d. If $b_{jk} \neq 0$ is not found, stop the Gaussian elimination.

When finishing the Gaussian elimination, the element of X_m is computed in the reversed order, only when $b_{ii} = 1$, from

$$x_i = y_i - \sum_{j=i+1}^{n} b_{ij}y_j \quad \text{for } (i = n, \ n-1, \ ..., \ 1) \tag{2.76}$$

where n is the maximum value of i with $b_{ii} = 1$.

Based on the recorded order change in X_m, each element of X_m can be mapped to the initial flux linkage $\lambda(0)$ in equation (2.67) for each AC voltage source to derive the additional voltage component.

Application Example

An example of the field-circuit coupling application is a 50 Hz, 125 MVA non-salient synchronous generator, as shown in Figure 2.25. The 2-pole field winding is excited by 1.3 kA DC current. The coupling circuit for three-phase armature windings is shown in Figure 2.26. The generator is operated at full load, with power angle of 35.7 electric degrees.

The simulation results with AF model are compared with those without using AF model. Figures 2.27 and 2.28 show the simulated current and torque waveforms without using AF model, respectively. The DC current components as shown in Figure 2.27 are still significant at time = 2 s, which is 100 periods. The DC current components generate a still magnetic field in the machine, which, interacting with the rotating field, produces alternating torque. The torque waveform profile in Figure 2.28 shows that it still needs a long time for the alternating torque component to decay to zero.

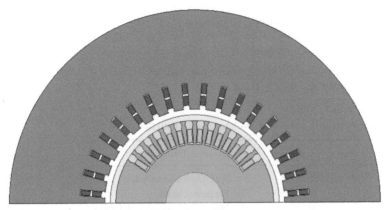

Figure 2.25   The 2D FEA model of the 125 MVA non-salient synchronous generator.

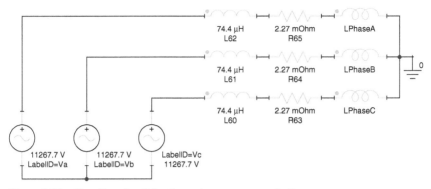

Figure 2.26   Coupling circuit for three-phase armature windings.

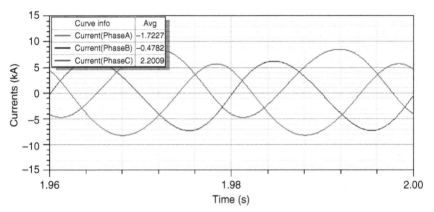

Figure 2.27   Simulated three-phase current waveforms without using AF algorithm.

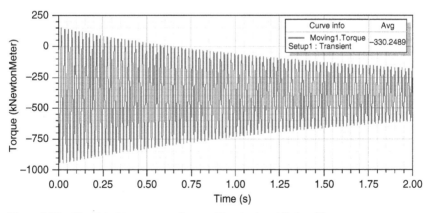

Figure 2.28   Simulated torque waveforms without using AF algorithm.

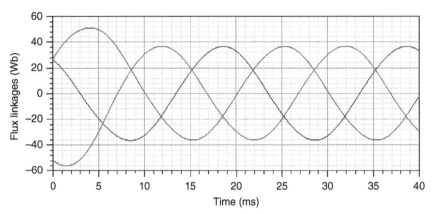

Figure 2.29   Simulated three-phase flux linkage waveforms with the AF model.

Figure 2.26 shows that there are only two loops in the coupling circuit, therefore, when using the AF model, one of the three AC voltage sources with AF model is dependent, and only the other two AC voltage source need to include the effects of the three-phase initial flux linkages. With the contribution of the additional voltage components in all three AC voltage sources, the simulated flux linkage waveforms reach the steady state within only two periods, as shown in Figure 2.29. The current and torque waveforms are shown in Figures 2.30 and 2.31, respectively.

2.4   HIGH PERFORMANCE COMPUTING—TIME DOMAIN DECOMPOSITION

Transient electromagnetic field simulation allows analysis of dynamic behaviors with nonlinear materials, permanent magnets, and induced eddy currents under various excitations. The transient process normally involves many time steps in a sequential

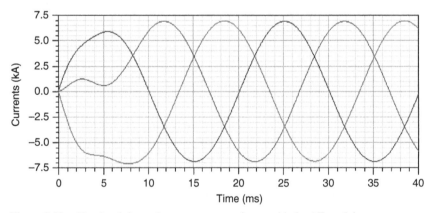

Figure 2.30   Simulated three-phase current waveforms with the AF model.

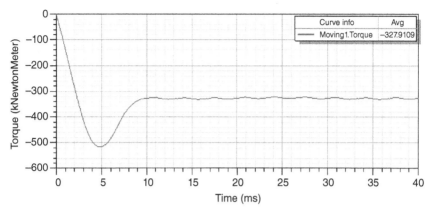

Figure 2.31   Simulated torque waveforms with the AF model.

fashion to calculate saturation, eddy currents, and rotor movement in time and space. Because transient electromagnetic analysis requires the computation of many time steps, the process is slow. In many cases, it is a huge computational undertaking to characterize an electric machine to steady-state operation. Sometimes it can take days or weeks to complete.

The basic procedure of transient simulation includes the spatial and temporal discretization of the physical equations. The finite element method (FEM) is widely used in engineering practice because using the irregular grids it can represent complicated geometry [33]. The FEM discretization produces a set of matrix differential equations. The typical temporal discretization includes backward Euler, Crank–Nicolson, and theta-method. Because of the nonlinearity, the matrices generally are dependent on the solution vectors, so an iteration method such as Newton–Raphson method should be used to solve these nonlinear matrix equations [7]. Namely, the nonlinear matrix equations are linearized for each nonlinear iteration. The linearized matrix equations may be solved by either a direct or iterative matrix solver. The transient simulation usually is time-consuming since it requires $N_t \times N_e$ number of matrix solving, where N_t is the number of time steps and N_e is the average number of nonlinear iterations. Provided that an algorithm (or method) can be made parallel, parallel computing can cut down the simulation time for the transient problems significantly. For example, parallel computing can be applied to the matrix solving at each time step. This can improve performance, but it is not always possible to make full use of all the parallel cores because the scaling is limited by various factors, such as that not all code executions are parallelable and there is communication overhead between the cores.

In order to achieve better parallel scalability, a new high performance computing technology is developed in terms of the domain decomposition along time-axis, namely TDM, to solve all time steps (or a subdivision of all time steps) simultaneously, instead of solving a transient problem sequentially one time step by one time step [34]. The transient simulations have a great range of engineering applications to simulate transient behavior of electromechanical devices. The spatial discretization of dynamical equations produces a semi-discrete form as

$$S(x,t)x(t) + \frac{d}{dt}[T(x,t)x(t)] = f(x,t) + \frac{d}{dt}[w(x,t)] \tag{2.77}$$

In the above, $S(x, t)$ and $T(x, t)$ are matrices, $f(x, t)$ and $w(x, t)$ are excitation vectors, and $x(t)$ is solution vector representing some physical quantity, such as the magnetic field, elastic displacement. Note that $S(x, t)$ and $T(x, t)$ are dependent on $x(t)$ to reflect the nonlinearity of the system. Depending on the physical system, not all of the above matrix or excitation vectors are dependent on $x(t)$. For example, only matrix $S(x, t)$ depends on $x(t)$ in transient FEA of electric machines based on the **A**-φ formulation [35].

For practical applications, we introduce two TDM models: one is the Periodic TDM Model for steady-state simulation based on solving all time steps in one period of time simultaneously, if the nature of physical problem is periodical; the other is the General TDM Model for general transient applications without the constraint of periodicity. It is based on solving a set of sequential subdivisions of all time steps, where all time steps in each subdivision are being solved simultaneously. Details are described later on how to achieve high parallel computing efficiency.

2.4.1 Periodic TDM Model

The periodic model is used to simulate the steady-state behavior of a system with a time periodic input. For this model, the initial condition is not required and in fact is not known a priori, and the solution and excitation vectors satisfy the periodic condition. That is,

$$x(t) = x(t+\tau), \quad f(t) = f(t+\tau), \quad w(t) = w(t+T) \tag{2.78}$$

hold with τ the period of the system. The semi-discrete form (equation 2.77) can be further discretized by applying the backward Euler method,

$$S_i x_i + \frac{T_i x_i - T_{i-1} x_{i-1}}{\Delta t} = f_i + \frac{w_i - w_{i-1}}{\Delta t} \tag{2.79}$$

Here the subscript i denotes the value of a quantity at time point t_i, for example, $x_i = x(t_i)$.

Although for simplicity and easy explanation, only the backward Euler method is used here for the temporal discretization, this is not a limitation to the method and other temporal discretization such as Crank–Nicolson or theta-method can also be applied. Based on the Newton–Raphson nonlinear iteration algorithm and using equation (2.79), the equation (2.77) becomes the following linearized matrix equations

$$
\begin{bmatrix}
K_1 & 0 & \cdots & 0 & M_n \\
M_1 & K_2 & \cdots & 0 & 0 \\
0 & M_2 & \ddots & \vdots & \vdots \\
\vdots & \vdots & \ddots & K_{n-1} & 0 \\
0 & 0 & \cdots & M_{n-1} & K_n
\end{bmatrix}
\begin{bmatrix}
\Delta x_1 \\
\Delta x_2 \\
\vdots \\
\Delta x_{n-1} \\
\Delta x_n
\end{bmatrix}
=
\begin{bmatrix}
b_1 \\
b_2 \\
\vdots \\
b_{n-1} \\
b_n
\end{bmatrix}
\tag{2.80}
$$

In the above, $K_i = \Delta t S_i' + T_i'$, and $M_i = -T_i'$, here S_i' and T_i' are the Jacobian matrices, Δx_i is the increment of solution vector during nonlinear iterations, and b_i is the residual vector in nonlinear iterations.

Let

$$
A =
\begin{bmatrix}
K_1 & 0 & \cdots & 0 & M_n \\
M_1 & K_2 & \cdots & 0 & 0 \\
0 & M_2 & \ddots & \vdots & \vdots \\
\vdots & \vdots & \ddots & K_{n-1} & 0 \\
0 & 0 & \cdots & M_{n-1} & K_n
\end{bmatrix}
\tag{2.81}
$$

The equation (2.80) can be written in concise form as

$$
A \, \Delta x = b
\tag{2.82}
$$

If the total unknowns per time step is m, the matrix of A is $(m \times n)$ by $(m \times n)$ size (associated with total unknowns for n time steps). Because of the block form of equation (2.80), it can be efficiently solved with proper parallel computing algorithm with either a direct solver or an iterative solver.

Dedicated Direct Solver
In order to solve the equation (2.82) using direct solver, it requires to inverse the matrix A. The computational cost of the inversion of the matrix A for such a huge size can be formidable, but fortunately can be significantly reduced by appropriately applying the Woodbury formula [36].

First, split the block matrix A into two parts

$$
A = \tilde{A} + \left(U \cdot V^T \right)
\tag{2.83}
$$

The part $\tilde{A}$ is defined as

$$
\tilde{A} = \begin{bmatrix}
K_1 & 0 & \cdots & 0 & 0 \\
M_1 & K_2 & \cdots & 0 & 0 \\
0 & M_2 & \ddots & \vdots & \vdots \\
\vdots & \vdots & \ddots & K_{n-1} & 0 \\
0 & 0 & \cdots & M_{n-1} & K_n
\end{bmatrix}
\tag{2.84}
$$

The other part is the product of two block vectors U and V^T which is defined as

$$
U = \begin{bmatrix}
u_1 \\
0 \\
\vdots \\
0 \\
0
\end{bmatrix}, V^T = \begin{bmatrix} 0 & 0 & \cdots & 0 & v_n \end{bmatrix}
\tag{2.85}
$$

where u_1 and v_n are sub-block column vectors with m (total unknowns per time step) by p. Here p is associated with the coupling with the last time step in sub-block matrix M_n. This coupling sub-block exists only for those unknowns associated with the windings of voltage source and electrical vector potential unknown **T** representing induced eddy current. Normally p is pretty small, for example, $p = 3$ if there are 3 voltage windings without induced eddy current. In addition, $\tilde{A}$ is a lower block triangular matrix, which means that its inversion is much easier than that of original matrix A.

It follows that using Woodbury matrix identity

$$
\left(\tilde{A} + UV^T \right)^{-1} = \tilde{A}^{-1} - \tilde{A}^{-1} U \left(I + V^T \tilde{A}^{-1} U \right)^{-1} V^T \tilde{A},
\tag{2.86}
$$

where I is an identity matrix.

We can avoid directly computing the inverse of A, which might be very costly to do, instead, compute the inverse of $(I + V^T \tilde{A}^{-1} U)$. The rank of $(I + V^T \tilde{A}^{-1} U)$ is only p (a p-by-p dense matrix) and the cost of the inverse is normally much cheaper than that for the original matrix A which is a $(m \times n)$ by $(m \times n)$ matrix with m as the total unknown per time step and n as the total time steps.

In addition to the advantages of using Woodbury matrix identity that we only need to work on the inverse of $(I + V^T \tilde{A}^{-1} U)$ with much smaller rank and that $\tilde{A}$ is a lower block triangular matrix whose inversion is much easier than that of A, the most attractive advantage is the inverse of each sub-block $(K_1, K_2, ..., K_n)$ can be independently done by distributed parallel computing. This very nicely fits TDM's objective.

Dedicated Iterative Solver

From the above discussion, clearly it is computationally very efficient to use the dedicated direct solver with the help of Woodbury matrix identity, if the column number p of sub-block column vectors u_1 and v_n in equation (2.85) is reasonably small. However, if the solution domain involves induced eddy current, the coupling sub-block matrix M_n normally becomes much denser, which leads to much big column number p. In such a case, the computational cost of using the above discussed direct solver will significantly increase. As a more efficient alternative solution, it is desirable to use iterative solver with a proper preconditioner.

An iterative solver with a preconditioner, such as the generalized minimal residual method (GMRES) or the biconjugate gradient method (BiCG) can be used to solve equation (2.82). A preconditioner is an approximation to the matrix A such that it can accelerate the convergence of iterations of the iterative solver [37, 38]. There are many ways to construct a preconditioner. For example, reference [35] proposed a preconditioner based on incomplete LU factorizations of sub-matrices, but the convergence is very slow. In order to accelerate the convergences, more efficient preconditioners are exploited here.

Preconditioners can be categorized into two types: non-overlapping preconditioners and overlapping preconditioners. The difference of them can be appreciated by the following example with $n = 6$, that is, six time steps. In such a case,

$$
A = \begin{bmatrix}
K_1 & & & & & M_6 \\
M_1 & K_2 & & & & \\
 & M_2 & K_3 & & & \\
 & & M_3 & K_4 & & \\
 & & & M_4 & K_5 & \\
 & & & & M_5 & K_6
\end{bmatrix}
\tag{2.87}
$$

Let C denote the preconditioner. For the case of two subdomains with domain 1 and domain 2, Jacobi preconditioner takes the form

$$
C_J = \begin{bmatrix}
K_1 & 0 & 0 & 0 & 0 & 0 \\
M_1 & K_2 & 0 & 0 & 0 & 0 \\
0 & M_2 & K_3 & 0 & 0 & 0 \\
0 & 0 & 0 & K_4 & 0 & 0 \\
0 & 0 & 0 & M_4 & K_5 & 0 \\
0 & 0 & 0 & 0 & M_5 & K_6
\end{bmatrix}
\tag{2.88}
$$

The two subdomains of the preconditioner are decoupled, so it can be parallelizable. However, the convergence of the iterative solver using Jacobi preconditioner is

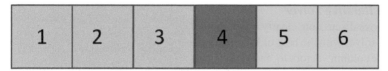

Figure 2.32   Two subdomains are overlapped.

slow. A better choice is to use Gauss–Seidel preconditioner, which is the lower triangular part of the matrix A, which takes the form

$$
C_{GS} =
\begin{bmatrix}
K_1 & 0 & 0 & 0 & 0 & 0 \\
M_1 & K_2 & 0 & 0 & 0 & 0 \\
0 & M_2 & K_3 & 0 & 0 & 0 \\
0 & 0 & M_3 & K_4 & 0 & 0 \\
0 & 0 & 0 & M_4 & K_5 & 0 \\
0 & 0 & 0 & 0 & M_5 & K_6
\end{bmatrix}
\tag{2.89}
$$

Gauss–Seidel preconditioner can significantly improve the convergence of the iterative solver, but with the cost of less parallelization. With the trade-off between parallelism and convergence, the additive Schwarz methods may be a better choice. The additive Schwarz methods can be conceived as generalization of Jacobi preconditioner. Unlike the Jacobi preconditioner, the subdomains may be overlapped. For example, the domain is composed of two subdomains as illustrated in Figure 2.32, that is, domain 1 ($K_1, M_1, K_2, M_2, K_3, M_3, K_4, M_4$) and domain 2 ($K_4, M_4, K_5, M_5, K_6, M_6$). The overlapped region is time step 4 (K_4, M_4).

2.4.2   General TDM Model

For General TDM Model, the solution vector $x(t)$ does not need to satisfy the periodic condition; thus, the initial condition is required. In addition, for most practical applications, the steady state of a device is of most interest. Thus, it will normally take several cycles of excitations to reach steady state. Due to constrains of hardware (available cores and memories), it might be not possible to solve all time steps altogether simultaneously. As a result, the gist of the general TDM model is to divide the entire nonlinear transient simulation into several subdivisions along the time-axis and for each subdivision, solve all time steps simultaneously. In such a case, the solution of the last time step in the current subdivision will be used as the initial condition for the subsequent subdivision.

For each subdivision, one needs to solve a block matrix at every nonlinear iteration

$$
A\Delta x = b
\tag{2.90}
$$

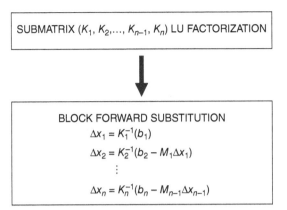

Figure 2.33   Direct block triangular matrix solver.

with

$$A = \begin{bmatrix} K_1 & 0 & \cdots & 0 & 0 \\ M_1 & K_2 & \cdots & 0 & 0 \\ 0 & M_2 & \ddots & \vdots & \vdots \\ \vdots & \vdots & \ddots & K_{n-1} & 0 \\ 0 & 0 & \cdots & M_{n-1} & K_n \end{bmatrix} \qquad (2.91)$$

As illustrated in Figure 2.33, it is noted that the inverse of each sub-block $(K_1, K_2, \ldots, K_n)$ can be independently done by distributed parallel computing; while the computation cost for block forward substitution is very small. Therefore, the block direct solver is exclusively used for the solution of the General TDM Model.

Clearly, the problem for Periodic TDM Model can be considered as a special case of that for General TDM Model. This means if a problem can be solved using Periodic TDM Model, it can also be solved using General TDM Model. Of course, computational efficiency is normally different. The opposite is not true. That is, if a problem can be solved using General TDM Model, it may not be solved using Periodic TDM model. In order to use Periodic TDM Model for transient simulation of electromechanical devices, all sources and induced eddy current in conducting region should have the same time period. For example, an induction machine cannot be solved by using Periodic TDM Model because the time period of currents in the stator winding is different from the time period of induced eddy current in the rotor bars.

2.4.3   Nonlinear Iteration

Figure 2.34 is the flowchart of the nonlinear iteration algorithm in the case of using the TDM for the transient simulation. The Newton–Raphson technique, due to its

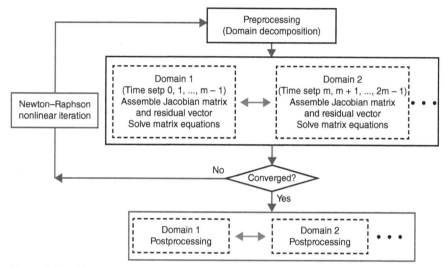

Figure 2.34   Flowchart of nonlinear iteration for the time decomposition method.

quadratic convergence characteristics, is a desirable choice. However, in some applications, such as the numerical simulation of electromechanical devices, the nonlinear iterations may converge at a very slow rate, or oscillate, or even diverge, especially with the new challenge that convergence has to be managed in a much bigger scope, that is, many time steps together, rather than examined by individual time step. To this end, certain techniques have been introduced to cope with the challenge.

For non-TDM transient, the solutions at previous time step can be used to estimate the initial nonlinear operating point for the first nonlinear iteration at each time step to speed up the convergence. But in TDM case, this scheme is no longer applicable. In such a case, a better choice is to start from linear portion of a nonlinear BH curve to have a monotonic convergence. At the same time, due to the significant difference in source excitation, position, etc., among different time steps, it is desirable to compute the individual nonlinear residual for each time step by proper scaling based on the normalized quantity in the right hand of the equations associated with individual time step [7].

In addition, because the relaxation factor which minimizes the square of the 2-norm of the residual with each iteration can be used to improve nonlinear convergence, the combined use of local relaxation factor associated with individual time step and the global relaxation factor associated with entire time steps will greatly improve the convergence behavior. At the beginning, the global relaxation factor is employed, but after certain iterations, the local relaxation factor will be further used. The local relaxation factor is determined based on the global relaxation factor and the convergence behavior indicated by the updated nonlinear residual of individual time step. The basic concept is that if the solution for a certain time step has reached a prescribed convergence criterion, a smaller relaxation should be used to avoid undesirable disturbance to the convergence of whole system.

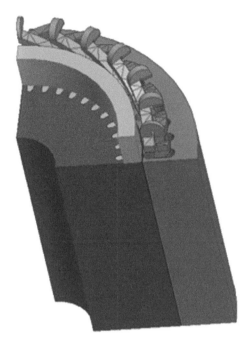

Figure 2.35   FEA solution domain of 1/8 geometry of 5.5 kW, 4-pole three phase induction motor.

2.4.4   Application Examples

The first example is a 5.5 kW, 4-pole three phase induction motor. The motor has 48 stator slots and 44 rotor bars as shown in Figure 2.35. The number of elements and unknowns are 131,475 and 248,634, respectively. Since the time period of currents in the stator winding is different from the time period of induced eddy current in the rotor bars, this problem has to be solved using General Transient TDM Model. Due to the constrain of available hardware resources (cores and memories), it normally has to be divided into several subdivisions along time-axis.

For this application, the number of total time steps is 640 to cover five electrical cycles. Table 2.5 shows the HPC performance of using MPI-based TDM against

TABLE 2.5   HPC performance of TDM for induction machine simulation

Number of MPI process	Number of subdivisions	Total simulation time (hours) (speedup)
8	80	135.3 (0.97)
16	40	77.1 (1.7)
32	20	43.5 (3.01)
64	10	23.3 (5.62)
128	5	12.9 (10.16)

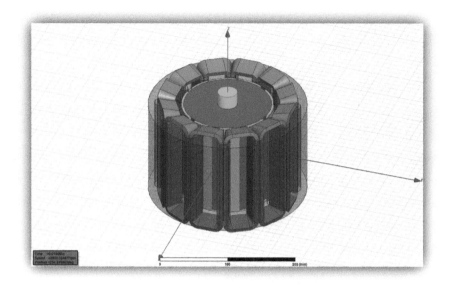

Figure 2.36   8 kW, 10-pole three phase BLDC motor.

non-TDM OpenMP with the use of 8 cores, since for OpenMP, the scalability beyond 8 cores has normally become saturated. The total simulation time of OpenMP is 131 hours, which is used as reference for comparison.

The second example is an 8 kW, 10-pole three phase BLDC motor as shown in Figure 2.36. The number of elements and unknowns are 226,312 and 460,045, respectively. This problem can be solved by either Periodical TDM model or General Transient TDM model. For the sake of illustration, Periodical TDM is used. In such a case, total 200 time steps over one period will be solved instantaneously.

Table 2.6 shows the HPC performance of TDM with different number of MPI processes against non-TDM OpenMP using 8 cores. The total simulation time of OpenMP is 16 hours 28 mins, which is used as reference for comparison.

In fact, the architecture of TDM is able to support two levels of parallelization at the same time. One is higher level with the distributed memory parallelization based

TABLE 2.6   HPC performance of TDM for BLDC motor simulation

Number of MPI process	Sequential time steps per MPI process	Cores for second level parallelization using OpenMP	Total simulation time (hours—mins) (speedup)
16	13	8	1 hour 31 mins (10.9)
32	7	4	1 hour 3 mins (15.8)
64	4	2	54 mins (18.2)
128	2	1	49 mins (20.1)
200	1	1	38 mins (26)

on MPI as the foundation for TDM. The other is lower level with shared memory parallelization based on OpenMP which is still applicable to the parallel computing inside each distributed MPI process. The number of cores used the lower level OpenMP is indicated in the third column. Please note if the distributed load is unbalanced, the actual computation time is determined by the process with the heaviest load. For example, for the first scenario with 16 MPI processes, for total 200 time steps, 8 MPI processes are assigned with 13 sequential time steps each and the other 8 MPI processes are assigned with 12 sequential time steps each. The actual time is determined by the MPI processes with 13 sequential time steps.

Also please note that for the first four scenarios, each scenario uses the same number of cores, that is, 128 cores, but computational efficiency is different. Clearly, MPI process associated with TDM is normally computationally more efficient than the use of OpenMP. Therefore, for practical applications, as long as the memory resource permits, it is advantageous to consider using more cores first on distributed MPI process as priority compared with assigning cores for the second level OpenMp.

2.5   REDUCED ORDER MODELING

In many electrical machine applications, system designers want to know the transient response to various electrical and mechanical inputs. Although FEA can simulate this transient process accurately, it usually requires very long processing time. In such numerical simulation of electrical machines, in which the magnetic field distribution inside electrical machines is not the main concern, reduced order modeling is an efficient technique. A key advantage of reduced order modeling is its capability for dramatically reducing the computational cost of numerical simulations, while maintaining a sufficient accuracy for the concerned performance from the engineering point of view. To this end, many researchers have applied reduced order model (ROM) to a lot of engineering problems such as circuit designs, hardware-in-the-loop controls.

Most ROMs of electrical machines are based on the electromechanical coupled state equations with some linear or nonlinear electrical and mechanical parameters. These parameters could be derived from traditional magnetic equivalent circuit (MEC) [39], or computed from FEA [40, 41]. Such kind of model can be referred as parameter-based ROM. The accuracy of a parameter-based ROM depends on the assumption conditions under which the state equations are derived. For example, the state equations using the $dq0$ system are derived at the assumption that the d- and q-axis inductance parameters are independent of the rotor position, therefore, the simulation results are accurate only when spatial field harmonics are negligible [42]. One advantage of the parameter-based ROM is it can handle problems with eddy currents.

An ROM can be also directly based on a look-up table which is computed from FEA at a series of sampling points of winding currents and rotor positions [43]. The outputs of the look-up table could be flux linkages of all windings, as well as rotor torque. The discrete output data at sampling points are transformed to continuous functions using multidimensional cubic spline interpolation and periodic extrapolation for rotor positions. Such kind of models can be referred as look-up-table based ROM, which is also referred as equivalent circuit extraction (ECE) model in this section. At the sampling points, the simulation results from a look-up-table based ROM are exactly the same as those from FEA. Therefore, the accuracy of a look-up-table based ROM could be very high as long as sampling points are dense enough. However, look-up-table based ROMs are normally not applicable to the problems with eddy currents.

In this section, we will introduce some advanced algorithms to create look-up-table based ROMs, and some efficient techniques to reduce computational time at specified density of the sampling points.

2.5.1 Sweep Strategy

An electrical machine may work at various excitations via single or poly-phase windings at various positions of moving part. By using field-circuit coupling for transient problems, the excitation sweeps, as well as the position sweep, can be separated into several individual sweep modules. Each module can be implemented as a model component in the circuit. The excitation and position sweeps of an electrical machine can be specified by one or more sweep modules in circuit schematic.

Single-Winding Current Sweep

The module for single-winding current sweep is abbreviated as single-winding module. In this module, we just need to specify the winding name and a list of sampling points for current sweep. This module can be used individually to create the ROM model of a single-phase inductor, or combining with other such modules to setup current sweeps for two or more windings to create the ROM model of a poly-phase inductor. This module can also be used to setup the current sweep for the field winding in a field excited synchronous machine.

Three-Phase Winding Current Sweep

The module for three-phase winding current sweeps is abbreviated as three-phase module. In this module, we need to set the three-phase winding names in the order of the positive sequence, and a list of sampling points for current sweep. We may also need to select a sweep type.

In a three-phase system, the sum of the three-phase currents is related to the zero-sequence component. In many cases, the zero-sequence three-phase currents just produce leakage flux. The leakage inductance related to the zero-sequence currents is usually linear. Therefore, from the sweep point of view, only two sweeps are independent. To get balanced three-phase currents, we introduce following three sweep types based on two independent sweeps.

i. (i_d, i_q) sweep

The (i_d, i_q) sweep is based on the transformation between the *abc* and *dq*0 systems. There are some formulations for *abc* to *dq*0 transformation, such as Park

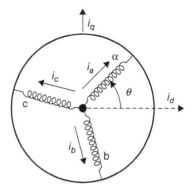

Figure 2.37   Three-phase *abc* axes and *dq* axes.

transformation and the power invariant transformation [42]. Based on the *d*- and *q*-axis directions as shown in Figure 2.37, we define the transformation between *abc* and *dq0* as

$$
\begin{bmatrix} i_a \\ i_b \\ i_c \end{bmatrix} = \begin{bmatrix} \cos(\theta) & \sin(\theta) & 1 \\ \cos\left(\theta - \frac{2\pi}{3}\right) & \sin\left(\theta - \frac{2\pi}{3}\right) & 1 \\ \cos\left(\theta - \frac{4\pi}{3}\right) & \sin\left(\theta - \frac{4\pi}{3}\right) & 1 \end{bmatrix} \cdot \begin{bmatrix} i_d \\ i_q \\ i_0 \end{bmatrix} = \mathbf{C} \cdot \begin{bmatrix} i_d \\ i_q \\ i_0 \end{bmatrix} \tag{2.92}
$$

In the sweeping process, (i_d, i_q) are set as sweeping variables, and i_0 is set to be zero. Since the three-phase currents are obtained based on the rotor position θ, the (i_d, i_q) sweep should be combined with rotor position sweep.

ii. (i_α, i_β) sweep

The (i_α, i_β) sweep is based on the transformation between the three-phase *abc* system and the two-phase $\alpha\beta0$ system. The transformation between *abc* and $\alpha\beta0$ is

$$
\begin{bmatrix} i_a \\ i_b \\ i_c \end{bmatrix} = \begin{bmatrix} 1 & 0 & 1 \\ \cos\left(\frac{2\pi}{3}\right) & \sin\left(\frac{2\pi}{3}\right) & 1 \\ \cos\left(\frac{4\pi}{3}\right) & \sin\left(\frac{4\pi}{3}\right) & 1 \end{bmatrix} \begin{bmatrix} i_\alpha \\ i_\beta \\ i_0 \end{bmatrix} \tag{2.93}
$$

In the sweeping process, (i_α, i_β) are set as sweeping variables, and i_0 is set to be zero.

iii. (i_m, β) sweep

For (i_m, β) sweep, the three-phase currents are given by

$$\begin{cases} i_a = i_m \cos(\beta) \\ i_b = i_m \cos(\beta - \frac{2\pi}{3}) \\ i_c = i_m \cos(\beta - \frac{4\pi}{3}) \end{cases} \tag{2.94}$$

In above three sweep types, current sweeps can be specified by a list of sampling points from 0 to a positive limited value. Required negative sampling points are obtained by symmetric extension from the positive part. In (i_m, β) sweep, number of intervals for β sweep, from 0 to 360 electrical degrees, is required from specification.

Rotational Position Sweep

The module for rotational position sweep is abbreviated as rotational module. In this module, we need to specify the maximum sweep angle in electrical degrees and the number of sweeping intervals. Since the rotor position in FEA is usually defined in mechanical degrees, we need also to specify the number of poles of the machine to transfer the position from electrical degrees to mechanical degrees. The initial rotor position in FEA is supposed to be located where the d-axis is aligned with the a-axis (the axis of phase a).

Combined with a three-phase module, we can setup excitation and position sweeps for three-phase PM machines. With an additional single-winding module, we can setup sweeps for three-phase field excited synchronous machines.

Translational Position Sweep

The module of translational (linear motion) position sweep is abbreviated as translational module. In this module, we need to specify the maximum sweep range in length unit, and the number of sweep intervals for the sweep range. There are two motion types for linear motion, one is limited motion which is used for actuators, and the other is periodic motion which is used for linear machines.

Combined with a three-phase module, we can setup sweeps for three-phase linear PM machines. Combined with a single-winding module, we can setup sweeps for actuators.

Transformer Excitation Current Sweep

In a poly-phase transformer, the primary current can be decomposed into two components in each phase: the excitation current component and the secondary current component. In each phase, the secondary current component in the primary winding is balanced by the current in the secondary winding. The resultant effect of this balance is to produce leakage flux. The leakage inductance related to this leakage flux is linear. Therefore, it is not necessary to sweep the secondary current, and only the excitation current is required to be swept for each phase.

The module of transformer excitation current sweep is abbreviated as transformer module. In this module, we need to specify winding names for all primary windings and relevant secondary windings. We need also to specify the turn ratio of the secondary windings to the primary windings. More than one secondary windings for each phase are allowed, but the turn ratios for all secondary windings must be specified.

The primary windings specified in transformer module must be defined in a three-phase module, or single-winding module, for excitation current sweep. In such a case, the (i_d, i_q) sweep type in the three-phase module is invalid. For a three-limb three-phase transformer, we can use the three-phase module to setup current sweeps because the zero-sequence flux will go from the top yoke to the bottom yoke via outer air space so that the zero-sequence inductance is linear. However, for a five-limb three-phase transformer, we have to use three single-winding modules for three-phase current sweeps because the zero-sequence flux will have closed path through the top and bottom yokes and two side limbs, and the zero-sequence inductance is nonlinear.

2.5.2   Look-Up Table Processing

Based on the combination of various modules, the look-up table is updated when field simulation is carried out for each sweep. When field simulations for all sweeps are finished, we obtain a virgin look-up table, which can be further processed to produce a full look-up table for ROM creation.

Output Types of Look-Up Table

For three-phase windings specified in three-phase module, the default output type for flux linkages is three-phase abc system. When the three-phase module is combined with a rotational module, and (i_d, i_q) sweep type is used, we can select one of the following two output types for the full look-up table:

 i. $(\lambda_d, \lambda_q, \lambda_0)$: flux linkages in $dq0$ system;

 ii. (L_d, L_q, λ_m): dq inductance and d-axis flux linkage by PM.

The flux linkages from FEA simulation are in abc system. When output type is $(\lambda_d, \lambda_q, \lambda_0)$, we need to complete the following transformation

$$\begin{bmatrix} \lambda_d \\ \lambda_q \\ \lambda_0 \end{bmatrix} = \mathbf{C}^{-1} \begin{bmatrix} \lambda_a \\ \lambda_b \\ \lambda_c \end{bmatrix} \tag{2.95}$$

where

$$\mathbf{C}^{-1} = \frac{2}{3} \begin{bmatrix} \cos(\theta) & \cos(\theta - \frac{2\pi}{3}) & \cos(\theta - \frac{4\pi}{3}) \\ \sin(\theta) & \sin(\theta - \frac{2\pi}{3}) & \sin(\theta - \frac{4\pi}{3}) \\ \frac{1}{2} & \frac{1}{2} & \frac{1}{2} \end{bmatrix} \tag{2.96}$$

For (L_d, L_q, λ_m) output type, according to the inductance matrix provided by FEA

$$\mathbf{L}_{abc} = \begin{bmatrix} L_{aa} & L_{ab} & L_{ac} \\ L_{ab} & L_{bb} & L_{bc} \\ L_{ac} & L_{ba} & L_{cc} \end{bmatrix} \tag{2.97}$$

the values of L_d and L_q can be obtained from the diagonal elements of the $dq0$ inductance matrix

$$\mathbf{L}_{dq0} = \mathbf{C}^{-1}\mathbf{L}_{abc}\mathbf{C} \tag{2.98}$$

and λ_m is obtained from

$$\lambda_m = \lambda_d - L_d i_d \tag{2.99}$$

In matrix $\mathbf{L}_{dq0}$, due to the slot and saturation effects, the non-diagonal elements may not be zero, and the 0 and q-axis flux linkages produced by PM may also not be zero. Ignoring these elements will cause the results based on (L_d, L_q, λ_m) output less accurate. In such a situation, it is meaningless to consider these outputs to be dependent on rotor position. Therefore, if rotor position is swept, average values over the range of the position sweep will be adopted for outputs, and the position sweep information is eliminated to reduce the look-up table size. A small size look-up table is very important for hardware-in-the-loop controls.

Determination of Zero-Sequence Inductance
In FEA results, the zero-sequence current does not have contribution to three-phase flux linkages, which can be taken into account by adding zero-sequence inductance in ROM.

Based on the FEA provided inductance matrix for the three-phase windings (equation 2.97), the zero-sequence inductance is computed from

$$L_0 = \frac{1}{3} \sum_{j=abc} \sum_{i=abc} L_{ij} \tag{2.100}$$

Due to slot effects, the result from equation (2.100) may not be very accurate. A better way is to use the average value obtained over entire current and position sweeps. After average zero-sequence inductance is obtained, its related flux linkages caused by the three-phase currents must be subtracted from the original three-phase flux linkages.

Extension of Look-Up Table
Since a rotor core normally has symmetric N and S poles, the flux linkage in one phase will be negatively repeatable in 180° (electrical) even though the three-phase windings are of the fractional-slot type. As long as the three-phase windings are symmetric, the flux linkage in one phase can be obtained from another phase by shifting

the waveform for 120°. Therefore, for the (i_m, β) sweep type, we just need to sweep the rotor position for 60° because the look-up table can be extended based on

$$
\begin{cases}
\lambda_a(i_m, \beta, \theta + 60°) = -\lambda_b(i_m, \beta - 60°, \theta) \\
\lambda_b(i_m, \beta, \theta + 60°) = -\lambda_c(i_m, \beta - 60°, \theta) \\
\lambda_c(i_m, \beta, \theta + 60°) = -\lambda_a(i_m, \beta - 60°, \theta) \\
T(i_m, \beta, \theta + 60°) = T(i_m, \beta - 60°, \theta)
\end{cases}
\tag{2.101}
$$

Generally, for the (i_d, i_q) sweep type, the rotor position needs to be swept for 120°, and the look-up table can be extended by

$$
\begin{cases}
\lambda_d(i_d, i_q, \theta + 60°) = \lambda_d(i_d, i_q, \theta) \\
\lambda_q(i_d, i_q, \theta + 60°) = \lambda_q(i_d, i_q, \theta) \\
\lambda_0(i_d, i_q, \theta + 120°) = \lambda_0(i_d, i_q, \theta) \\
T(i_d, i_q, \theta + 60°) = T(i_d, i_q, \theta)
\end{cases}
\tag{2.102}
$$

The zero-sequence flux linkage λ_0 in equation (2.102) includes triple harmonic components only. For odd triple (such as 3rd, 9th, …) harmonics, λ_0 satisfies

$$
\lambda_0(i_d, i_q, \theta + 60°) = -\lambda_0(i_d, i_q, \theta)
\tag{2.103}
$$

For even triple (such as 6th, 12th, …) harmonics, λ_0 satisfies

$$
\lambda_0(i_d, i_q, \theta + 60°) = \lambda_0(i_d, i_q, \theta)
\tag{2.104}
$$

Usually, in an electrical machine, if the resultant air-gap field is not distorted, all even harmonics are zero, and λ_0 includes only odd triple harmonics. In such cases, the rotor position just needs to be swept for 60°, and the look-up table extension for λ_0 can be obtained based on equation (2.103). However, if the resultant air-gap field is seriously distorted, such as in IPM machines, λ_0 may include both odd and even triple harmonics, therefore we need to sweep rotor position for 120°.

Processing for Skewed Core

If the stator or rotor core is skewed, its effects will not be included in 2D FEA results. Provided a rotor position sweep is involved, the skew effects can be considered by

$$
f(\theta) = \frac{1}{\delta} \int_{\theta - \delta/2}^{\theta + \delta/2} f'(\theta) d\theta
\tag{2.105}
$$

where δ is the skew angle specified in the rotational module, $f'(\theta)$ represents any one output of the look-up table, and $f(\theta)$ is the modified one.

Computation of Leakage Inductance for Transformer

Consider a single-phase transformer with one primary winding and m secondary windings. The inductance matrix from FEA will be

$$\mathbf{L} = \begin{bmatrix} L_{00} & L_{01} & \cdots & L_{0m} \\ L_{01} & L_{11} & \cdots & L_{1m} \\ \vdots & \vdots & \ddots & \vdots \\ L_{0m} & L_{1m} & \cdots & L_{mm} \end{bmatrix}, \tag{2.106}$$

where index 0 denotes the primary and indexes from 1 to m are for the secondary. Referring all parameters from the secondary windings to the primary winding, we get referred inductance matrix as

$$\mathbf{L}' = \begin{bmatrix} L_{00} & r_1 L_{01} & \cdots & r_m L_{0m} \\ r_1 L_{01} & r_1^2 L_{11} & \cdots & r_1 r_m L_{1m} \\ \vdots & \vdots & \ddots & \vdots \\ r_m L_{0m} & r_1 r_m L_{1m} & \cdots & r_m^2 L_{mm} \end{bmatrix}, \tag{2.107}$$

where coefficients r_1 to r_m are turn ratios for each secondary winding. If any element of $\mathbf{L}'$ is denoted as L'_{ij} ($i = 0, 1, \ldots, m; j = 0, 1, \ldots, m$), we can find the minimum mutual inductance as

$$M = \min(L'_{ij}) \ \text{for} \ (i = 0, 1, \ldots, m; \ j = i + 1, \ldots, m) \tag{2.108}$$

Then leakage inductance for the primary and all secondary windings will be

$$l_i = L'_{ii} - M \ \text{for} \ (i = 0, 1, \ldots, m) \tag{2.109}$$

Leakage inductances for other phases are computed in the same way. To get more accurate results, average values are preferred over all sampling points.

2.5.3   Circuit Model Creation

Once the full look-up table is obtained, a circuit model (ROM) can be created. According to the combination of different sweep modules, various circuit models can be derived. Below, we just introduce some models for typical applications.

Circuit Model for Rotating Machines

Based on the full look-up table, and the zero-sequence inductance L_0, a circuit model for three-phase PM rotating machines with (i_d, i_q) sweep type can be created as shown in Figure 2.38.

In Figure 2.38, the current transformation from abc to $dq0$ is given by

$$\begin{bmatrix} i_d \\ i_q \\ i_0 \end{bmatrix} = \mathbf{C}^{-1} \begin{bmatrix} i_a \\ i_b \\ i_c \end{bmatrix} \tag{2.110}$$

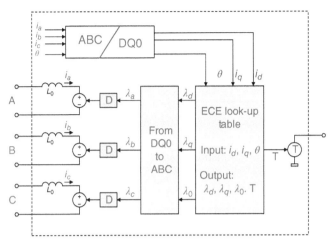

Figure 2.38   Circuit model for three-phase PM rotating machines.

and the flux linkage transformation from $dq0$ to abc is expressed as

$$\begin{bmatrix} \lambda_a \\ \lambda_b \\ \lambda_c \end{bmatrix} = \mathbf{C} \cdot \begin{bmatrix} \lambda_d \\ \lambda_q \\ \lambda_0 \end{bmatrix} \qquad (2.111)$$

For each phase, the transformation from flux linkage to induced voltage can be realized by a circuit in which a current source controlled by the flux linkage signal is connected in series with an inductor with unit-value inductance. The induced voltage can be represented by a voltage source controlled by the voltage across the inductor.

The signal across the torque component is speed. To get rotor position from speed, we need initial rotor position from model specification. The position signal can be obtained from the speed signal by an integrating circuit, in which a current source controlled by speed signal is connected in series with a capacitor with unit-value capacitance. The initial voltage of the capacitor is obtained from the model-specified initial position, and the voltage across the capacitor represents the position signal.

Circuit Model for Actuators
The circuit model for actuators is shown in Figure 2.39, where the mechanical *through* source is force, and the across source is position. The limit stop model "S" limits the position within the lower and upper limits.

Circuit Model for Transformers
The circuit model for three-phase transformers in terms of (i_α, i_β) sweep type is shown in Figure 2.40, where the primary and secondary resistances are obtained from model specifications. Since the look-up table outputs flux linkages directly in abc system, no further transformation is required.

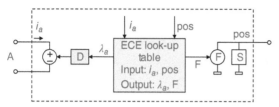

Figure 2.39   Circuit model for actuators.

In Figure 2.40, the current transformation from abc to $\alpha\beta0$ is

$$
\begin{bmatrix} i_\alpha \\ i_\beta \\ i_0 \end{bmatrix} = \frac{2}{3} \begin{bmatrix} 1 & \cos\left(\frac{2\pi}{3}\right) & \cos\left(\frac{4\pi}{3}\right) \\ 0 & \sin\left(\frac{2\pi}{3}\right) & \sin\left(\frac{4\pi}{3}\right) \\ \frac{1}{2} & \frac{1}{2} & \frac{1}{2} \end{bmatrix} \begin{bmatrix} i_a \\ i_b \\ i_c \end{bmatrix}
\tag{2.112}
$$

If (i_m, β) sweep type is used, the inputs of the look-up table will be

$$
\begin{cases} i_m = \sqrt{i_\alpha^2 + i_\beta^2} \\ \beta = \arctan(i_\beta, i_\alpha) \end{cases}
\tag{2.113}
$$

2.5.4   Application Examples

Two application examples are presented here: one is a 550 W, 8/6 pole switched reluctance motor (SRM), and the other is a 4-pole, 1500 rpm three-phase synchronous PM generator.

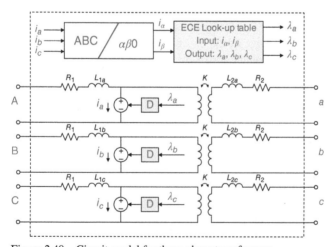

Figure 2.40   Circuit model for three-phase transformers.

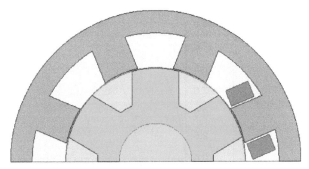

Figure 2.41   2D FEA model for the switched reluctance motor with only one-phase winding for current sweep.

For the SRM, the four-phase winding excitations are triggered by 80° (electrical) pulse signals with each successive phase lagging 90°. Hence, when a phase starts to be triggered, the previous phase has almost finished free wheeling. In such a case, the mutual effects between adjacent two phases are negligible. Therefore, only one phase current is swept combining with the rotor position sweep. The 2D FEA model for the sake of ROM creation is shown in Figure 2.41. The winding current has 11 uniform sampling points sweeping from 0 to 20 A, and the rotor position has 60 sampling points sweeping from 0 to 60° (mechanical) for one rotor pole. The total number of sampling points is 660.

The circuit connection for four separate one-phase ROMs of 8/6 pole SRM is shown in Figure 2.42, where the specified initial position for the first ROM is 0, and those for the other three ROMs increase 90° for each successive phase.

The SRM is analyzed by both circuit and FEA simulators with the same DC voltage excitations of 220 V at the same constant speed of 940 rpm. The simulated current and torque waveforms from both simulators are compared in Figures 2.43 and 2.44 , respectively.

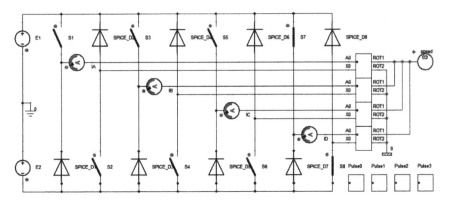

Figure 2.42   Circuit connection for four separate one-phase ROMs of 8/6 pole SRM.

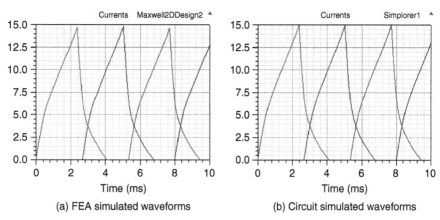

(a) FEA simulated waveforms (b) Circuit simulated waveforms

Figure 2.43   Simulated current waveforms of the switched reluctance motor for all phases. (a) FEA simulated waveforms; (b) circuit simulated waveforms.

The 1-pole FEA model of the 4-pole PM generator is shown in Figure 2.45. The (i_d, i_q) sweep type is used for the stator three-phase current sweep. Both i_d and i_q have five uniform sampling points sweeping from −20 A to 20 A. The rotor position has 24 sampling points sweeping from 0 to 120° (electrical). The total number of sampling points is 600.

The generator is simulated at no-load and load conditions by both FEA and circuit simulators. The simulated no-load induced voltage, load current, and load torque waveforms from both simulators are compared in Figures 2.46, 2.47, and 2.48, respectively.

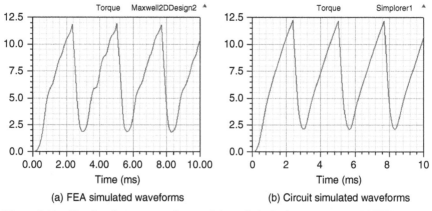

(a) FEA simulated waveforms (b) Circuit simulated waveforms

Figure 2.44   Simulated torque waveforms of the switched reluctance motor. (a) FEA simulated waveforms; (b) circuit simulated waveforms.

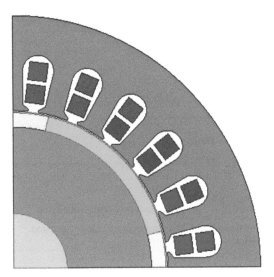

Figure 2.45   One-pole 2D FEA model for the 4-pole PM generator.

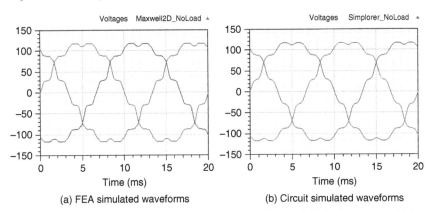

(a) FEA simulated waveforms   (b) Circuit simulated waveforms

Figure 2.46   Simulated no-load induced voltage waveforms of the PM generator. (a) FEA simulated waveforms; (b) circuit simulated waveforms.

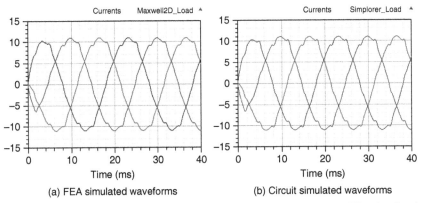

(a) FEA simulated waveforms   (b) Circuit simulated waveforms

Figure 2.47   Simulated load current waveforms of the PM generator. (a) FEA simulated waveforms; (b) circuit simulated waveforms.

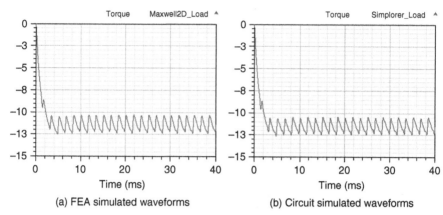

(a) FEA simulated waveforms

(b) Circuit simulated waveforms

Figure 2.48 Simulated load torque waveforms of the PM generator. (a) FEA simulated waveform; (b) circuit simulated waveform.

REFERENCES

[1] P. Zhou, D. Lin, W. N. Fu, B. Ionescu, and Z. J. Cendes, "A general co-simulation approach for coupled field-circuit problems," *IEEE Trans. Magn.*, vol. 42, no. 4, pp. 1051–1504, 2006.

[2] D. Lin, P. Zhou, Z. Badics, W. N. Fu, Q. M. Chen, and Z. J. Cendes, "A new nonlinear anisotropic model for soft magnetic materials," *IEEE Trans. Magn.*, vol. 42, no. 4, pp. 963–966, Apr. 2006.

[3] A. Mohammed and N. A. Demerdash, "An extremely fast technique for nonlinear three dimensional finite element magnetic field computations," *IEEE Trans. Magn.*, vol. 23, no. 5, pp. 3575–3577, Sep. 1987.

[4] L. Janicke and A. Kost, "Convergence properties of the Newton-Raphson method for nonlinear problems," *IEEE Trans. Magn.*, vol. 34, no. 5, pp. 2505–2508, Sep. 1998.

[5] T. Nakata, N. Takahashi, K. Fujiwara, N. Okamoto, and K. Muramatsu, "Improvements of convergence characteristics of Newton-Raphson method for nonlinear magnetic field analysis," *IEEE Trans. Magn.*, vol. 28, no. 2, pp. 1048–1051, Mar. 1992.

[6] J. O'Dwyer and T. O'Donnell, "Choosing the relaxation parameter for the solution of nonlinear magnetic field problems by Newton-Raphson method," *IEEE Trans. Magn.*, vol. 31, no. 3, pp. 1484–1487, 1995.

[7] P. Zhou, D. Lin, B. He, S. Kher, and Z. J. Cendes, "Strategies for accelerating non-linear convergence for T-Omega formulation," *IEEE Trans. Magn.*, vol. 46, pp. 3129–3132, 2010.

[8] P. Leonard, R. J. Hill-Cottingham, and D. Rodger, "3D finite element models and external circuits using the AΨ scheme with cuts," *IEEE Trans. Magn.*, vol. 30, no. 5, pp. 3220–3223, 1994.

[9] J. P. Webb, B. Forghani, and D. A. Lowther, "An approach to the solution of three-dimensional voltage driven and multiply connected eddy problems," *IEEE Trans. Magn.*, vol. 28, no. 2, pp. 1193–1196, 1992.

[10] F. Henrotte and K. Hameyer, "An algorithm to construct the discrete cohomology basis functions required for magnetic scalar potential formulations without cuts," *IEEE Trans. Magn.*, vol. 39, no. 3, pp. 1167–1170, 2003.

[11] G. Meuuier, F. Y. Le, and C. Guérin, "A nonlinear circuit coupled t-t_0-Φ formulation for solid conductors," *IEEE Trans. Magn.*, vol. 39, no. 3, pp. 1729–1732, 2003.

[12] Z. Ren, P. Zhou, and Z. J. Cendes, "Computation of current vector potential due to excitations in multiply connected conductors," in *Proc. ICEF Int. Conf.*, Tianjin, China, 2000, pp. 121–124.

[13] Z. Ren, "T-Ω; formulation for eddy-current problems in multiply connected regions," *IEEE Trans. Magn.*, vol. 38, no. 2, pp. 557–560, 2002.

[14] P. Zhou, Z. Badics, D. Lin, and Z. J. Cendes, "Nonlinear T-Ω; formulation including motion for multiply connected 3-D problems," *IEEE Trans. Magn.*, vol. 44, no. 6, pp. 718–721, Jun. 2008.

[15] C. Lu, P. Zhou, D. Lin, and D. Sun, "Multiply connected 3D transient problem with rigid motion associated with T-Ω; formulation," *IEEE Trans. Magn.*, vol. 50, Article#. 7011004, Feb. 2014.

[16] F. Piriou and A. Razek, "Finite element analysis in electromagnetic systems accounting for electrical circuit," *IEEE Trans. Magn.*, vol. 29, pp. 1669–1675, Mar. 1993.

[17] S. J. Salon, M. J. DeBortoli, and R. Palma, "Coupling of transient fields, circuits, and motion using finite element analysis," *J. Electromagn. Waves & Appl.*, vol. 4, pp. 1077–1108, Nov. 1990.

[18] P. Zhou, W. N. Fu, D. Lin, S. Stanton, and Z. J. Cendes, "Numerical modeling of magnetic devices," *IEEE Trans. Magn.*, vol. 40, pp. 1803–1809, Jul. 2004.

[19] I. A. Tsukerman, A. Konrad, G. Meunier, and J. C. Sabonnadiere, "Coupled field-circuit problems: Trends and accomplishments," *IEEE Trans. Magn.*, vol. 29, pp. 1701–1704, Mar. 1993.

[20] G. Bedrosian, "A new method for coupling finite element field solutions with external circuits and kinematics," *IEEE Trans. Magn.*, vol. 29, pp. 1664–1668, Mar. 1993.

[21] S. Kanerva, "Data transfer methodology between a FEA program and a system simulator," in *Proc. IEEE-ICEMS Fifth Int. Conf.*, 2001, pp. 1121–1124.

[22] P. Zhou, D. Lin, W. N. Fu, B. Ionescu, and Z. J. Cendes, "A general cosimulation approach for coupled field-circuit problems," *IEEE Trans. Magn.*, vol. 42, no. 4, pp. 1051–1054, Apr. 2006.

[23] P. Zhou, J. R. Brauer, S. Stanton, and Z. J. Cendes, "Dynamic modeling of universal motors," in *IEEE International Electric Machines and Drives Conference IEMDC '99*, Seattle, Washington, May 9–12, 1999, pp. 419–421.

[24] T. Matsuda, T. Moriyama, N. Konda, Y. Suzuki, and Y. Hashimoto, "Method for analyzing the commutation in small universal motors," *Proc. Inst. Elect. Eng., Electr. Power Appl., vol.* 142, no. 2, pp. 123–130, Mar. 1995.

[25] R. H. Wang and R. T. Walter, "Model of universal motor performance and brush commutation using finite element computed inductance and resistance matrices," *IEEE Trans. Energy Convers.*, vol. 15, no. 3, pp. 257–263, Sep. 2000.

[26] A. D. Gerlando and R. Perini, "Model of the commutation phenomena in a universal motor," *IEEE Trans. Energy Convers.*, vol. 21, no. 1, pp. 27–33, Mar. 2006.

[27] D. Lin, P. Zhou, W. N. Fu, B. Ionescu, and Z. J. Cendes, "A flexible approach for brush-commutation machine simulation," *IEEE Trans. Magn.*, vol. 44, no. 6, pp. 1542–1545, Jun. 2008.

[28] S. Ausserhofer, O. Bíró, and K. Preis, "An efficient harmonic balance method for nonlinear eddy-current problems," *IEEE Trans. Magn.*, vol. 43, no. 4, pp. 1229–1232, Apr. 2007.

[29] O. Bíró and K. Preis, "An efficient time domain method for nonlinear periodic eddy current problems," *IEEE Trans. Magn.*, vol. 42, no. 4, pp. 695–698, Apr. 2006.

[30] S. Nogawa, M. Kuwata, T. Sakura, K. Fujiwara, and N. Takahashi, "Examination of initial values of three-phase shunt reactor for fast 3-D steady state eddy current analysis," in *Proc. 11th Int. IGTE Symp. Numer. Field Calc. Elect. Eng.*, Graz, Austria, Sep. 2004, pp. 30–33.

[31] D. N. Dyck and P. J. Weicker, "Periodic steady-state solution of voltage-driven magnetic devices," *IEEE Trans. Magn.*, vol. 43, no. 4, pp. 1533–1536, Apr. 2007.

[32] S. A. Mousavi, C. Carrander, and G. Engdahl, "Comprehensive study on magnetization current harmonics of power transformers due to GICs," in *International Conference on Power Systems Transients (IPST 2013)*, Vancouver, Canada, Jul. 2013, pp. 18–20.

[33] O. C. Zienkiewicz, R. L. Taylor, and J. Z. Zhu, *The Finite Element Method: Its Basis and Fundamentals.* Butterworth-Heinemann, 2006.

[34] B. He, C. Lu, P. Zhou, D. Lin, and N. Chen, "Time domain decomposition method in transient," Patent serial no. 62209155.

[35] Y. Takahashi, T. Tokumasu, M. Fujita, T. Iwashita, H. Nakashima, S. Wakao, and K. Fujiwara, "Time-domain parallel finite-element method for fast magnetic field analysis of induction motors," *IEEE Trans. Magn.*, vol. 49, no. 5, pp. 2413–2416, May 2013.

[36] W. W. Hager, "Updating the inverse of a matrix," *SIAM Review*, vol. 31, no. 2, pp. 221–239, 1989.

[37] H. A. van der Vorst, *Iterative Krylov Methods for Large Linear Systems.* Cambridge University Press, 2003.

[38] B. Smith, P. Bjorstad, and W. Gropp, *Domain Decomposition Parallel Multilevel Methods for Elliptic Partial Differential Equations*. Cambridge University Press, 1996.

[39] S. A. Saied, K. Abbaszadeh, and M. Fadaie, "Reduced order model of developed magnetic equivalent circuit in electrical machine modeling," *IEEE Trans. Magn.*, vol. 46, no. 7, pp. 2649–2655, Jul. 2010.

[40] S.-M. Lee, S.-H. Lee, H.-S. Choi, and II.-H. Park, "Reduced modeling of eddy current-driven electromechanical system using conductor segmentation and circuit parameters extracted by FEA," *IEEE Trans. Magn.*, vol. 41, no. 5, pp. 1448–1451, May 2005.

[41] W. N. Fu, X. Zhang, and S. L. Ho, "A fast frequency-domain parameter extraction method using time-domain FEM," *IEEE Trans. Magn.*, vol. 50, no. 2, Article no. 7010604, Feb. 2014.

[42] N. N. Hancock, *Matrix Analysis of Electrical Machinery*, 2nd ed. New York: Pergamon, 1974.

[43] S. Lin, X. Li, T. Wu, L. Chow, Z. Tang, and S. Stanton, "Temperature dependent reduced order IPM motor model based on finite element analysis," in *Proc. IEEE Int. Electr. Mach. Drives Conf. (IEMDC)*, May 11–13, 2015, pp. 543–549.

CHAPTER **3**

MAGNETIC MATERIAL MODELING

3.1 SHAPE PRESERVING INTERPOLATION OF *B–H* CURVES

In finite element analysis (FEA) involving nonlinear magnetic materials, it is necessary to obtain a continuous mathematical representation of a finite set of discrete measured data for *B–H* curves. The methodology of such a mathematical representation falls into two categories: interpolation and approximation.

The term approximation refers to seeking a function best fitted to the given data set. It is not necessary for the approximated *B–H* curve, or the fitting curve, to pass through all given data points.

Interpolation is a special approximation with the interpolant passing through all specified data points. Shape preserving interpolation means the interpolant maintains the shape implied in the discrete *B–H* data, or more clearly, the shape (monotonicity and convexity) of the curve gained by joining the *B–H* data by straight line segments (which we call the "piecewise linear interpolant") [1].

A typical *B–H* curve and its derivative are shown in Figure 3.1.

Figure 3.1 shows that the B–H curve has positive convexity when $h < h_m$, and negative convexity when $h > h_m$. The convexity of a curve can be described by the monotonicity of its derivative. In general, the properties of the *B–H* curve shape can be described by its derivative as:

1. the derivative increases monotonically at first, decreases monotonically after it reaches its maximum value at h_m, then approaches a constant value of μ_0;
2. the derivative is greater than, or equal to, μ_0 for $h \in [0, \infty)$.

If the input *B–H* data set is given as (h_1, b_1), (h_2, b_2), ..., (h_n, b_n), at points $h_1 < h_2 < ... < h_n$, then according to the above properties, the input data set can be

Multiphysics Simulation by Design for Electrical Machines, Power Electronics, and Drives, First Edition.
Marius Rosu, Ping Zhou, Dingsheng Lin, Dan Ionel, Mircea Popescu, Frede Blaabjerg, Vandana Rallabandi, and David Staton.
© 2018 by The Institute of Electrical and Electronics Engineers, Inc. Published 2018 by John Wiley & Sons, Inc.

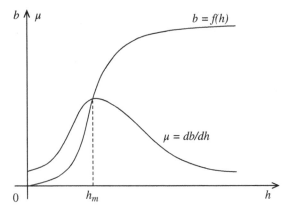

Figure 3.1   A typical *B–H* curve and its derivative.

validated at the very beginning by

$$\begin{cases} \delta_i \geq \mu_0 & i \in (1, 2, \ldots, n-1) \\ \delta_i \leq \delta_{i+1} & i \in (1, 2, \ldots, m-1) \\ \delta_i \geq \delta_{i+1} & i \in (m, \ldots, n-2) \end{cases} \tag{3.1}$$

where

$$\delta_i = (b_{i+1} - b_i)/(h_{i+1} - h_i) \; i \in (1, 2, \ldots, n-1) \tag{3.2}$$

and m is the index for δ_m, the maximum value of δ_i.

In many cases, due to some limitations of measurement conditions, the measured *B–H* data may not be saturated enough to reach the free-space permeability μ_0. Therefore, shape preserving extrapolation may also be required [2].

In the following sections, we will introduce some shape preserving interpolation (including extrapolation) schemes. The input *B–H* data are assumed to have been validated by (3.1).

3.1.1   Piecewise Linear Interpolation

The simplest shape preserving interpolation is the piecewise linear interpolation. The interpolant for h inside segment $[h_i, h_{i+1}]$, $i \in (1, 2, \ldots, n-1)$, is expressed as

$$f(h) = (1 - t)b_i + t b_{i+1} \tag{3.3}$$

where

$$t = (h - h_i)/(h_{i+1} - h_i) \tag{3.4}$$

which ranges from 0 to 1.

The derivative of interpolant is constant in each segment and is given by (3.2). The interpolant and its derivative are shown in Figure 3.2. The *dB/dH* curve

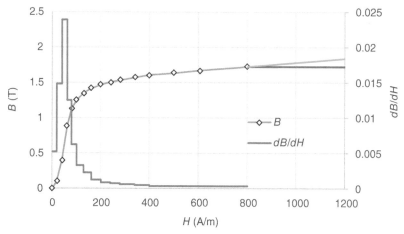

Figure 3.2   Piecewise linear interpolant and its derivative.

in Figure 3.2 shows that the derivative of the linear interpolant discontinues at each sampling point, which may cause the Newton–Raphson iteration in FEA not to converge to a smaller error.

When $h > h_n$, an extrapolation is required. According to the shape properties of *B–H* curves, when electrical steels get deeply saturated, the slope of the *B–H* curve approaches a slope equal to the free-space permeability μ_0. Therefore, an often-heard advice is to extrapolate the *B–H* curve with a slope equal to μ_0 whenever a point beyond the end of the *B–H* curve provided by the manufacturer is required (the red-line extrapolation in Figure 3.2). Another often-used approach is to extend the *B–H* curve with the slope of the last two points (the green-line extrapolation in Figure 3.2). Both approaches are feasible only when the relative permeability of the last data point provided is close to 1. Otherwise, besides causing large errors, the first extrapolation approach may also cause the Newton–Raphson iteration to diverge at large field.

In order for the Newton–Raphson iteration to converge to an error as small as the specified one, it is desired that the derivative of the interpolant continues in the whole region: $h \in [0, \infty)$, and the extrapolating curve smoothly approaches to the slope of μ_0. Therefore, some more advanced interpolating algorithms are preferred.

3.1.2   Cubic Spline Interpolation

The cubic spline interpolant for h inside $[h_i, h_{i+1}]$, $i \in (1, 2, \ldots, n)$, is expressed as [3]

$$f(h) = k_{00}b_i + k_{01}\Delta h_i u_i + k_{10}b_{i+1} + k_{11}\Delta h_i u_{i+1} \tag{3.5}$$

where

$$
\begin{cases}
k_{00} = 2t^3 - 3t^2 + 1 \\
k_{01} = t^3 - 2t^2 + t \\
k_{10} = -2t^3 + 3t^2 \\
k_{11} = t^3 - t^2
\end{cases}
\tag{3.6}
$$

$$
\Delta h_i = h_{i+1} - h_i
\tag{3.7}
$$

t is obtained from (3.4), and u_j, $i \in (1, 2, \dots, n)$, being determined in the pre-processing, are the derivatives of the interpolant at points h_i.

The key issue for the cubic spline interpolation is to determine the discrete derivatives u_i, $i \in (1, 2, \dots, n)$, in the pre-processing. There exist a lot of selections for u_i as long as the following conditions are satisfied:

1. $u_i \geq \mu_0$;
2. the discrete derivatives u_i increase monotonically for $i < m$, and decrease monotonically for $i > m$;
3. the derivative of interpolant in each segment is monotonic.

If u_i are selected to be between two values of δ_{i-1} and δ_i for $i \in (2, \dots, n-1)$, the above conditions (1) and (2) can be fulfilled for all internal points. Therefore,

$$
\begin{cases}
u_i = \alpha_i \delta_{i-1} + (1 - \alpha_i)\delta_i i \in (2, \dots, n-1) \\
u_1 = 2\delta_1 - u_2 \\
u_n = 2\delta_{n-1} - u_{n-1}
\end{cases}
\tag{3.8}
$$

where α_i satisfies

$$
0 \leq \alpha_i \leq 1
\tag{3.9}
$$

and can be initialized as 0.5. In (3.8), if u_1 or $u_n < \mu_0$, let u_1 or $u_n = \mu_0$. Then u_i will be modified to ensure that condition (3) is fulfilled by verifying the second derivatives of the interpolant in the segment.

In the derivative increasing region for i from 2 to $m - 1$, the second derivatives at two endpoints in all segments should not be negative. Based on (3.5)–(3.7), the second derivatives of the interpolant at two endpoints in the segment $[h_i, h_{i+1}]$ can be computed from

$$
\begin{cases}
g_{0,i} = f''(h_i) = -(4u_i + 2u_{i+1} - 6\delta_i)/\Delta h_i \\
g_{1,i} = f''(h_{i+1}) = (2u_i + 4u_{i+1} - 6\delta_i)/\Delta h_i
\end{cases}
\tag{3.10}
$$

In this segment, u_i and/or u_{i+1} will be modified to ensure both $g_{0,i} \geq 0$ and $g_{1,i} \geq 0$, retaining $g_{1,i-1} \geq 0$ (for the previous segment $[h_{i-1}, h_i]$).

First, u_i and/or u_{i+1} are modified to ensure $g_{0,i} \geq 0$. If $g_{0,i} < 0$, the selected u_i is too large. In such a case, we can decrease u_i by letting $g_{0,i} = 0$. Decreasing u_i will cause $g_{1,i-1}$ to decrease. If $g_{1,i-1} < 0$, which means u_i is over modified, then we can increase u_i by letting $g_{1,i-1} = 0$. Increasing u_i will cause $g_{0,i}$ to become negative again, but it has been improved, compared with the original situation. In such a case, we can modify u_{i+1} by letting $g_{0,i} = 0$.

Then, u_{i+1} is modified by letting $g_{1,i} = 0$ if $g_{1,i} < 0$. In such a case, $g_{0,i}$ is positive. Decreasing u_{i+1} will cause $g_{0,i}$ to decrease, but $g_{0,i}$ will not decrease to a negative value.

In the above two steps, condition (3.9) must always be satisfied.

Now, we will continue to work with the next segment until i reaches $m - 1$.

In a similar way, u_i can be modified to ensure that the second derivative of interpolant in each segment is not positive in the derivative decreasing region where $i > m$.

3.1.3   Quadratic Spline Interpolation

The quadratic spline interpolation is also based on the pre-selected discrete derivative set. For quadratic spline interpolation, the second derivative of interpolant is constant in each segment. In (3.10), let $g_{0,i} = g_{1,i}$, we obtain

$$0.5u_i + 0.5u_{i+1} = \delta_i \tag{3.11}$$

Equation (3.11) gives a constraint for each segment [4]. For n points of a discrete derivative set, there are $n - 1$ constraints. Therefore, only one derivative is able to be freely selected. In such a case, it is almost impossible to retain the basic properties of *B–H* curve by properly selecting the derivatives for each sampling point.

If we add one knot in the center of each segment, and denote the derivative of the center knot as u_{ic}, due to linear interpolation for derivatives in each half segment, the constraint (3.11) becomes

$$0.25u_i + 0.5u_{ic} + 0.25u_{i+1} = \delta_i \tag{3.12}$$

Then, the derivatives in all sampling points can be freely selected. Free selection of discrete derivatives makes it possible to preserve monotonicity and convexity for input data.

The field and derivative values of the center knot can be obtained by

$$\begin{cases} h_{ic} = 0.5(h_i + h_{i+1}) \\ u_{ic} = 2\delta_i - 0.5(u_i + u_{i+1}) \\ b_{ic} = b_i + 0.5(u_i + u_{ic})(h_{ic} - h_i) \end{cases} \tag{3.13}$$

It can be proved that if the discrete derivative set is selected based on the same algorithm described in Section 3.1.2, u_{ic} will be between (u_i, u_{i+1}), and the derivative of interpolant is monotonic in the segment.

Based on the discrete data at two endpoints and one center knot in each segment, we can get b from h, $h \in [h_i, h_{i+1}]$, as

$$b = \begin{cases} b_i + u_i(h - h_i) + 0.5g_{0,i}(h - h_i)^2 & h < h_{ic} \\ b_{ic} + u_{ic}(h - h_{ic}) + 0.5g_{1,i}(h - h_{ic})^2 & h \geq h_{ic} \end{cases} \tag{3.14}$$

where $g_{0,i}$ and $g_{1,i}$ are constant second derivatives of the interpolants in each half segment, and are given by

$$\begin{cases} g_{0,i} = (u_{ic} - u_i)/(h_{ic} - h_i) \\ g_{1,i} = (u_{i+1} - u_{ic})/(h_{i+1} - h_{ic}) \end{cases} \tag{3.15}$$

We can extend the B–H curve by quadratic extrapolation, or linear extrapolation for the derivative data (h_i, u_i), $i \in (1, 2, \ldots, n)$. One more point can be added at the end where the extended derivative curve reaches μ_0. Beyond the new last point, the derivative curve is extended with the constant μ_0. The data of the new last point can be assigned as

$$\begin{cases} h_{n+1} = h_n + (\mu_0 - u_n)/g_{1,n-1} \\ u_{n+1} = \mu_0 \\ b_{n+1} = b_n + 0.5(u_n + u_{n+1})(h_{n+1} - h_n) \end{cases} \tag{3.16}$$

The quadratic interpolant and its derivative based on the same sampling points for linear interpolation in Figure 3.2 are shown in Figure 3.3, and Figure 3.4 shows the extrapolation of the B–H curve and its derivative.

We can also get h from b, $b \in [b_i, b_{i+1}]$, as

$$h = \begin{cases} h_i + \dfrac{2(b - b_i)}{u_i + \sqrt{u_i^2 + 2g_{0,i}(b - b_i)}} & b < b_{ic} \\[3ex] h_{ic} + \dfrac{2(b - b_{ic})}{u_{ic} + \sqrt{u_{ic}^2 + 2g_{1,i}(b - b_{ic})}} & b \geq b_{ic} \end{cases} \tag{3.17}$$

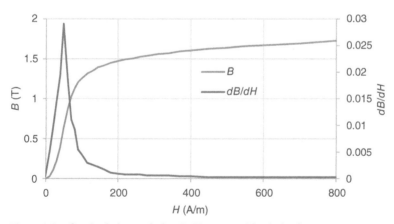

Figure 3.3   Quadratic interpolation B–H curve and its derivative.

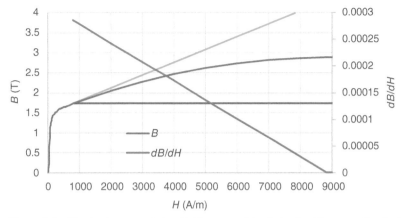

Figure 3.4   Quadratic extrapolation B–H curve and its derivative (the green line for linear extrapolation, the red line for μ_0 extrapolation).

Equation (3.17) is obtained from $h = f^{-1}(b)$, which means both (3.14) and (3.17) are based on the same B–H and derivative curves. This property is very important when it is required to get b from h, and get h from b, alternately in an iteration algorithm.

3.2   NONLINEAR ANISOTROPIC MODEL

Anisotropic behavior is often encountered not only in the grain oriented but also in the laminated grain non-oriented magnetic cores in electromagnetic devices.

One commonly used model for anisotropic material is called the elliptical model [5, 6]. This model derives the permeability in the principal directions directly from the relevant B–H curves based on the magnitude of the applied magnetic field. If a two-dimensional (2D) magnetic field with constant magnitude is applied in different directions, the permeability of each principal direction is constant and, therefore, the vector of the flux density traces an ellipse. However, as reported in [7], this model does not provide good accuracy.

The co-energy model for grain-oriented steels [8] utilizes the stored co-energy density over the H_x–H_y plane based on four B–H curves. Two of the curves correspond to the applied field in the two principal directions of the 2D plane, and the other two curves correspond to the B components in tangent and normal directions to the applied field at 55° to the rolling direction. Whenever necessary, the flux density B can be recovered as the gradient of the co-energy density. For three-dimensional (3D) anisotropic problems, it is difficult to define the B–H curves at 55°.

Another method for modeling grain-oriented steel incorporates a far larger number of magnetization curves [7]. The steel is cut into strips in 10° steps with respect to the rolling direction and tested to get B–H curves in 10 different directions. These curves are then used in numerical computation. Obviously, this method has the following drawbacks: (i) the difference in the directions of **B** and **H** cannot be considered, therefore, the method is only suitable for problems where the B components orthogonal to **H** are negligible; (ii) it is not convenient to obtain B–H curves in so many directions; (iii) it is difficult to expand the method to 3D nonlinear anisotropic problems.

In this section, based on the elliptical model, an improved anisotropic model is introduced for nonlinear anisotropic soft materials [9]. This model requires only B–H curves in the principal directions—two for 2D or three for 3D. Such curves are normally provided directly by the manufacturers. Furthermore, lamination effects are also represented in the model.

3.2.1 Improved Anisotropic Model

If a magnetic field is applied in a principal direction in a nonlinear anisotropic soft material with negligible hysteresis, the directions of **H** and **B** are the same and the magnitude of the applied magnetic field **H** and of the flux density **B** simply follow the relevant B–H curve. However, when **H** is applied in any other direction, **B** and **H** are no longer parallel.

Based on the fact that the application of the same magnitude of **H** in different principal directions will cause different levels of magnetic saturation, the proposed model refers all components of **H** to the principal directions. An equivalent magnitude of the magnetic field is introduced in each principle direction to consider both anisotropy and the cross effects of different directions due to nonlinearity. The equivalent magnitudes in the x, y, and z directions are computed as

$$\begin{cases} H_e^x = \sqrt{H_x^2 + (k_{xy}H_y)^2 + (k_{xz}H_z)^2} \\ H_e^y = \sqrt{(k_{yx}H_x)^2 + H_y^2 + (k_{yz}H_z)^2} \\ H_e^z = \sqrt{(k_{zx}H_x)^2 + (k_{zy}H_y)^2 + H_z^2} \end{cases} \tag{3.18}$$

The referring coefficients in (3.18) are defined as

$$\begin{cases} k_{xy} = 1/k_{yx} = W_m^y/W_m^x \\ k_{xz} = 1/k_{zx} = W_m^z/W_m^x \ , \\ k_{yz} = 1/k_{zy} = W_m^z/W_m^y \end{cases} \tag{3.19}$$

where the magnetic co-energy densities W_m^x, W_m^y, and W_m^z are obtained from the W–H curves at the magnitude H_m of the applied field **H** as shown in Figure 3.5. The W–H curve in Figure 3.5 is obtained by integrating the corresponding B–H curve.

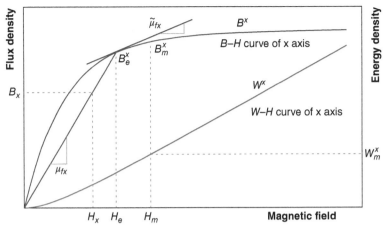

Figure 3.5   Permeability and W–H curve for axis x (or y or z by replacing x by y or z).

The permeability in each principal direction is obtained in terms of the corresponding equivalent magnetic field magnitudes, H_e^x, H_e^y, or H_e^z, and is expressed as

$$
\begin{cases}
\mu_{fx} = B^x\left(H_e^x\right)/H_e^x \\
\mu_{fy} = B^y\left(H_e^y\right)/H_e^y \\
\mu_{fz} = B^z\left(H_e^z\right)/H_e^z
\end{cases}
\tag{3.20}
$$

where $B^x(\cdot)$, $B^y(\cdot)$, and $B^z(\cdot)$ represent the B–H curves in x-, y-, and z-axes, respectively.

Finally, the flux density $\mathbf{B}$ is calculated from

$$
\mathbf{B} = [\mu] \cdot \mathbf{H},
\tag{3.21}
$$

where $[\mu]$ is the permeability tensor

$$
[\mu] =
\begin{bmatrix}
\mu_{fx} & 0 & 0 \\
0 & \mu_{fy} & 0 \\
0 & 0 & \mu_{fz}
\end{bmatrix}
\tag{3.22}
$$

If all referring coefficients in (3.19) are assigned to be 1, all equivalent field magnitudes in (3.18) are the same, and the proposed model is degenerated into the elliptical model [5, 6].

3.2.2   Lamination Effects

For the sake of generality, this improved model can also take lamination effects into account. Assume the steel is laminated in the z direction, that is, the magnetic paths in the x- and y-axes are performed by the parallel connection of the iron and insulation parts of laminations, and that in the z-axis is in series connection. Therefore, the field intensities of the iron part in the x and y directions are the same as the applied ones, but that in the z direction should be scaled by a factor $k_{\mu z} < 1$ due to the series connection. Hence, the equivalent magnitudes of the field intensities for the lamination iron, in (3.20), are computed from

$$
\begin{cases}
H_e^x = \sqrt{H_x^2 + (k_{xy}H_y)^2 + (k_{xz}k_{\mu z}H_z)^2} \\[2mm]
H_e^y = \sqrt{(k_{yx}H_x)^2 + H_y^2 + (k_{yz}k_{\mu z}H_z)^2} \quad , \\[2mm]
H_e^z = \sqrt{(k_{zx}H_x)^2 + (k_{zy}H_y)^2 + (k_{\mu z}H_z)^2}
\end{cases} \tag{3.23}
$$

where

$$
k_{\mu z} = \frac{\mu_0}{(1 - k_{\text{lam}})\mu_{fz} + k_{\text{lam}}\mu_0} \tag{3.24}
$$

with k_{lam} being the stacking factor of the lamination, which is 1 for solid steels.

The permeability tensor considering the lamination effects for (3.21) is

$$
[\mu] = \begin{bmatrix} \mu_x & 0 & 0 \\ 0 & \mu_y & 0 \\ 0 & 0 & \mu_z \end{bmatrix}, \tag{3.25}
$$

where

$$
\begin{cases}
\mu_x = k_{\text{lam}}\mu_{fx} + (1 - k_{\text{lam}})\mu_0 \\
\mu_y = k_{\text{lam}}\mu_{fy} + (1 - k_{\text{lam}})\mu_0 \quad . \\
\mu_z = k_{\mu z}\mu_{fz}
\end{cases} \tag{3.26}
$$

Equation (3.25) becomes (3.22) when $k_{\text{lam}} = 1$. Therefore, (3.25) is valid for both solid and laminated cores.

The result of (3.24) depends on (3.20), which in turn depends on (3.23) and (3.24). Therefore, a simple iterative process is required. Figure 3.6 shows the flowchart of the iterative process. In general, this iterative process completes in just a couple of iterations with the initial value of $k_{\mu z} = 1.0$.

A non-oriented electrical steel, Armco M-15 with stacking factor of 0.95, is simulated with a magnetic field applied in directions 0° (x-axis), 55° (relative to the x-axis toward the z-axis) and 90° (z-axis, i.e., the lamination direction). Since the material is isotropic, all referring coefficients from (3.19) are unity. For each applied field in the above three different directions, the flux density can be computed from (3.21), where $[\mu]$ is determined by (3.25) and (3.26). In this way, B–H curves for the above three directions can be accurately obtained, as shown in Figure 3.7. Since **B** and **H** are not in the same direction when field **H** is applied in 55°, curves labeled with 55T and 55N denote the **B** components in the tangent and normal directions of **H**, respectively.

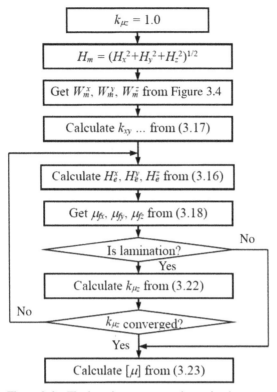

Figure 3.6   The iterative process to determine $k_{\mu z}$.

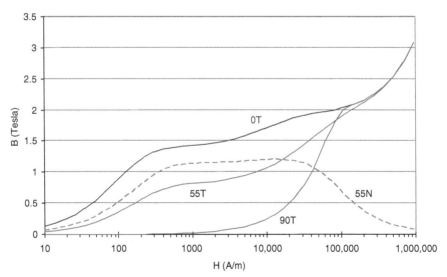

Figure 3.7   Numerical experiment results applying a magnetic field at 0°, 55°, and 90° in M-15 non-oriented steel. T and N denote the tangential and normal directions, respectively.

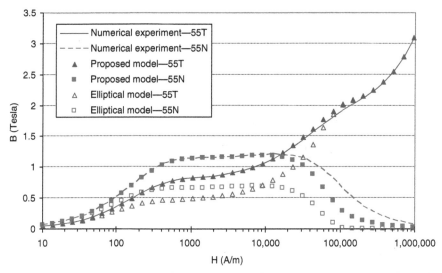

Figure 3.8 Comparison between the numerical experiment and the computed results by the proposed model and the elliptical model.

Now, if the stacked core is treated as an anisotropic solid core with $k_{\text{lam}} = 1$, the B–H curves simulated from the stacked core, shown in Figure 3.7 can be used as measured data for the anisotropic solid core. This method is regarded as numerical experiment. Using the B–H curves in the principal directions at 0° and 90° from the numeric experiment as the input data, we can simulate the B–H curves at 55° by the proposed anisotropic model described in Section 3.2.1. Thus, the simulated B–H curves at 55° can be validated by those from the numeric experiment.

The purpose of this simulation is to see how well the computed results match experimental data at the critical direction 55° [7, 8]. The computed results of the **B** components in the tangent and normal directions of **H** are compared with the numerical experiment data in Figure 3.8. For the sake of comparison, the computed results obtained by the elliptical model in [5, 6] are also given in Figure 3.8. It can be seen that the proposed model is much more accurate than the elliptical model.

3.2.3 Jacobian Matrix

The nonlinear anisotropic model introduced in the previous sections can be incorporated into a finite element Newton–Raphson iteration scheme where the flux density is computed from

$$B = B_0 + [\tilde{\mu}] \cdot (H - H_0), \tag{3.27}$$

where B_0 and H_0 denote the previous field solution, and $[\tilde{\mu}]$ is the Jacobian matrix expressed as

$$[\tilde{\mu}] = dB/dH = [\Delta\tilde{\mu}] + [\mu] \tag{3.28}$$

Note that $[\Delta\tilde{\mu}]$ in (3.28) is in general a full tensor. For laminations with anisotropic steel, $[\Delta\tilde{\mu}]$ can be derived as

$$[\Delta\tilde{\mu}] = A \cdot \begin{bmatrix} \Delta\tilde{\mu}_{fx} \cdot a_x b_x & \Delta\tilde{\mu}_{fx} \cdot a_x b_y & \Delta\tilde{\mu}_{fx} \cdot a_x b_z \\ \Delta\tilde{\mu}_{fy} \cdot a_y b_x & \Delta\tilde{\mu}_{fy} \cdot a_y b_y & \Delta\tilde{\mu}_{fy} \cdot a_y b_z \\ 0 & 0 & 0 \end{bmatrix}$$

$$+ C \cdot \begin{bmatrix} \Delta\tilde{\mu}_{fx} \cdot c_x d_x & \Delta\tilde{\mu}_{fx} \cdot c_x d_y & \Delta\tilde{\mu}_{fx} \cdot c_x d_z \\ \Delta\tilde{\mu}_{fy} \cdot c_y d_x & \Delta\tilde{\mu}_{fy} \cdot c_y d_y & \Delta\tilde{\mu}_{fy} \cdot c_y d_z \\ \Delta\tilde{\mu}_{fz} \cdot c_z d_x & \Delta\tilde{\mu}_{fz} \cdot c_z d_y & \Delta\tilde{\mu}_{fz} \cdot c_z d_z \end{bmatrix}, \quad (3.29)$$

where

$$\begin{cases} \Delta\tilde{\mu}_{fx} = \tilde{\mu}_{fx} - \mu_{fx} = \partial B^x/\partial H_e^x - B_e^x/H_e^x \\ \Delta\tilde{\mu}_{fy} = \tilde{\mu}_{fy} - \mu_{fy} = \partial B^y/\partial H_e^y - B_e^y/H_e^y \\ \Delta\tilde{\mu}_{fz} = \tilde{\mu}_{fz} - \mu_{fz} = \partial B^z/\partial H_e^z - B_e^z/H_e^z \end{cases} \quad (3.30)$$

$$\begin{cases} a_x = \dfrac{H_x}{k_{xz}}\dfrac{\partial k_{xz}}{\partial H_m} = -\dfrac{H_x}{k_{zx}}\dfrac{\partial k_{zx}}{\partial H_m} = \dfrac{H_x}{k_{xz}}\dfrac{B_m^z - k_{xz}B_m^x}{W_m^x} \\ a_y = \dfrac{H_y}{k_{yz}}\dfrac{\partial k_{yz}}{\partial H_m} = -\dfrac{H_y}{k_{zy}}\dfrac{\partial k_{zy}}{\partial H_m} = \dfrac{H_y}{k_{yz}}\dfrac{B_m^z - k_{yz}B_m^y}{W_m^y} \end{cases} \quad (3.31)$$

$$\begin{cases} b_x = \partial H_m/\partial H_x = H_x/H_m \\ b_y = \partial H_m/\partial H_y = H_y/H_m \\ b_z = \partial H_m/\partial H_z = H_z/H_m \end{cases} \quad (3.32)$$

$$\begin{cases} c_x = H_x/H_e^z \\ c_y = H_y/H_e^z \\ c_z = k_{muz}{}^2 H_z/H_e^z \end{cases} \quad (3.33)$$

$$\begin{cases} d_x = c_x/k_{xz}{}^2 - b_x k_{ac} \\ d_y = c_y/k_{yz}{}^2 - b_y k_{ac} \\ d_z = c_z - b_z k_{ac} \end{cases} \quad (3.34)$$

and

$$k_{ac} = a_x c_x/k_{xz}{}^2 + a_y c_y/k_{yz}{}^2 \quad (3.35)$$

$$A = k_{\text{lam}} \quad (3.36)$$

$$C = \frac{k_{\text{lam}}}{1 + c_z^2(\Delta \tilde{\mu}_{fz}/\mu_0)(1 - k_{\text{lam}})/k_{\mu z}} \tag{3.37}$$

For laminations with isotropic steel, where all referring coefficients from (3.19) are one, we have

$$a_x = a_y = k_{ac} = 0 \tag{3.38}$$

$$d_x = c_x, \quad d_y = c_y, \quad d_z = c_z \tag{3.39}$$

Therefore,

$$[\Delta \tilde{\mu}] = C \cdot \Delta \tilde{\mu}_f \begin{bmatrix} c_x^2 & c_x c_y & c_x c_z \\ c_x c_y & c_y^2 & c_y c_z \\ c_x c_z & c_y c_z & c_z^2 \end{bmatrix} \tag{3.40}$$

It can be seen from (3.28) and (3.40) that for laminations with isotropic steel, $[\tilde{\mu}]$ is a symmetric tensor.

The model is implemented in a 3D FEA solver using the T–Ω formulation. Stable convergence of the Newton–Raphson algorithm is ensured by utilizing an adaptive relaxation factor.

3.2.4 Case Study: Synchronous Reluctance Motor

Figure 3.9 displays a 400 W 4-pole synchronous reluctance motor with axially laminated anisotropic rotors. To make material property assignment easy, it is convenient to use a cylindrical coordinate system fixed to the rotor. In such a case, the

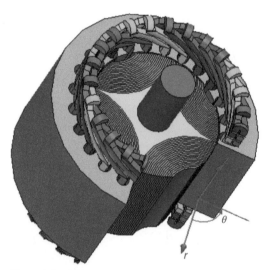

Figure 3.9   The structure of a 400 W 4-pole synchronous reluctance motor with axially laminated anisotropic rotor.

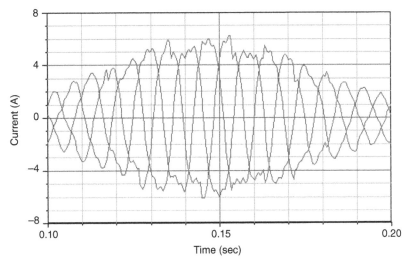

Figure 3.10   Three-phase current waveforms when the rotor rotates with a slip of 150 rpm.

lamination direction is in the radial direction r shown in Figure 3.9. Due to periodic and symmetry conditions, only one-fourth of the arrangement is analyzed.

The air-gap torque at rated voltage varying with the torque angle is computed based on the slip method. The torque angle is defined as the phase difference between the phasors of the phase voltage and the phase induced voltage. The torque angle can be adjusted by a small slip speed.

A three-phase 110 V, 60 Hz voltage source is applied to the three-phase winding of the motor, creating a rotating field with synchronous speed of 1800 rpm. The rotor is forced to rotate at a speed of 1650 rpm, a little lower than the synchronous speed. The performance of the motor is analyzed in one slip period, from 0 to 0.2 s, corresponding to 5 Hz slip frequency, or 150 rpm slip speed.

The computed three-phase currents are shown in Figure 3.10. The current amplitude varies between 2 A and 6 A, corresponding to rotor positions when the d-axis is aligned with or perpendicular to the axis of the stator rotating field. The d-axis and q-axis reactances, x_d and x_q, are derived from the ratio of the winding flux linkage to the winding current at these two special positions. Table 3.1 shows the computed results of x_d and x_q. Notice that x_d and x_q are nonlinear, and the values are only valid for a certain condition. The magnetic field distribution for the x_d computation is shown in Figure 3.11.

The computed air-gap torque varying with the torque angle is shown in Figure 3.12. We can see that the torque is not zero when the torque angle is zero. This

TABLE 3.1   Computed d-axis and q-axis reactances

x_d	x_q
77.0 ohms	25.6 ohms

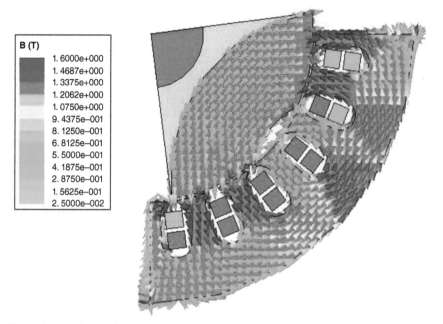

Figure 3.11 The *d*-axis magnetic field distribution.

is due to the effects of the stator winding resistance. We can also observe that the computed torque curve includes harmonics due to the effect of slots during slip operation. In reality, the curve is smoother because the average torque should be derived over one period at synchronous speed. The real torque curve can be obtained by field

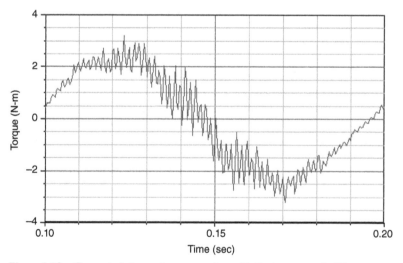

Figure 3.12   Computed air-gap torque varying with the torque angle. The torque angle varies from 0 to 180 electric degrees as time changes from 0.1 to 0.2 s.

analysis at synchronous speed for various torque angles. Of course, this will significantly increase the computation time.

3.3 DYNAMIC CORE LOSS ANALYSIS

In order to predict core loss in transient FEA in the design of magnetic power devices such as inductors, transformers, and electric machines, it is required to calculate core loss in the time domain. While approaches exist for loss computation in power devices in the frequency domain, an appropriate method for core loss computation in the time domain remains unclear.

In the frequency domain, loss separation is widely used with problems involving magnetic laminations. Loss separation breaks the total core loss into static hysteresis loss P_h, classical eddy current loss P_c, and excess loss P_e [10]

$$\begin{aligned} P_v &= P_h + P_c + P_e \\ &= k_h f B_m^{\beta} + k_c (f B_m)^2 + k_e (f B_m)^{1.5} \end{aligned} \qquad (3.41)$$

Given the coefficients k_h, k_c, k_e, and the parameter β, the total core loss per unit volume P_v in the frequency domain can be calculated in terms of peak magnetic flux density B_m and frequency f. When this approach is applied to the time domain, the computation of the eddy current loss and the excess loss is straightforward. However, the computation of hysteresis loss is still difficult.

With ferrite materials, a well-known empirical approach proposed by Steinmetz a century ago is normally used

$$P_v = C_m f^{\alpha} B_m^{\beta}, \qquad (3.42)$$

where C_m, α, and β are empirical parameters obtained from experimental measurement under sinusoidal excitation [11]. This approach is also suitable in the frequency domain only.

In the following sections, a time domain dynamic hysteresis model is developed for soft magnetic and power ferrite materials based on the idea of an equivalent elliptical loop (EEL). First, the model is introduced as "post processing," that is, the effects of the core loss on the transient magnetic field are not taken into account [12]. This model is able to consider the effects of minor loops and predicts instantaneous hysteresis loss with good accuracy. In addition, the required parameters in the model are the same as those required in the frequency domain approaches (3.41) and (3.42). These parameters are either directly available from manufacturers or can be easily extracted from standard loss curves under sinusoidal excitation.

Then, the effects of the lamination core loss on the 3D transient magnetic field are considered by introducing an additional field component in lamination regions when using the T-Ω method [13]. This additional field component is derived from the total instantaneous core loss including the static hysteresis loss and excess loss. An iteration algorithm is developed to ensure that the iteration of solving this additional field component does not affect the convergence of the Newton–Raphson iteration.

The effects of classical eddy current loss can be taken into account by directly solving the field equation with modified permeability in the lamination region [14].

3.3.1 Dynamic Core Loss Model

A typical value of hysteresis loss parameter β in (3.41) is 2. In this case, the magnetic field H in a static hysteresis loop can be decomposed into two components: a reversible component H_{rev} and an irreversible component H_{irr}. As a result, hysteresis loss can be computed by

$$
\begin{aligned}
P_h &= \frac{1}{T} \int_0^T (H_{rev} + H_{irr}) \frac{dB}{dt} dt \\
&= \frac{1}{T} \int_0^T H_{irr} \frac{dB}{dt} dt
\end{aligned}
\tag{3.43}
$$

The reversible component can be directly obtained from the normal B–H curve without considering a hysteresis loop. In fact, H_{rev} is related to the reactive power in the material and H_{irr} is associated with the hysteresis loss. Consequently, the instantaneous hysteresis loss is

$$
p_h(t) = H_{irr} \frac{dB}{dt}
\tag{3.44}
$$

Equation (3.44) indicates that the key to computing $p_h(t)$ is the procedure used to obtain H_{irr}. Figure 3.13 defines an EEL in which H_{irr} is evaluated by tracing an elliptical loop having the same area as that of the original hysteresis loop.

The ellipse in Figure 3.13 can be described as

$$
\begin{cases}
B = B_m \sin(\theta) \\
H_{irr} = H_m \cos(\theta)
\end{cases}
\tag{3.45}
$$

where B_m is directly obtained from a historical record of the flux density and H_m is determined by requiring that the core loss calculated in the time domain must be the same as that obtained in the frequency domain under the same sinusoidal excitation. From (3.43) and (3.45), the time-average hysteresis loss with sinusoidal excitation is

$$
\begin{aligned}
P_h &= H_m \cdot B_m \cdot 2\pi f \cdot \frac{1}{T} \int_0^T \cos^2(2\pi f t) dt \\
&= H_m \cdot B_m \cdot \pi f
\end{aligned}
\tag{3.46}
$$

Let (3.46) be equal to the frequency domain solution of $P_h = k_h f B_m^2$. Then we have

$$
H_m = \frac{1}{\pi} k_h \cdot B_m
\tag{3.47}
$$

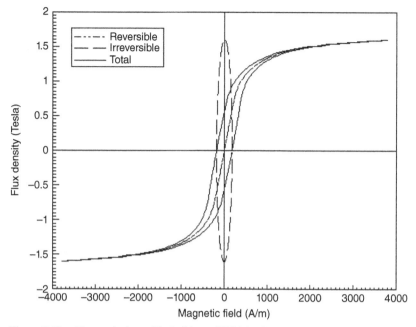

Figure 3.13 The equivalent elliptical loop (EEL) having the same area as the original hysteresis loop.

Thus,

$$H_{\text{irr}} = \frac{1}{\pi}k_h \cdot B_m \cos(\theta) \tag{3.48}$$

With a similar procedure, the eddy current loss and excess loss in the time domain can be expressed as, respectively,

$$p_c(t) = \frac{1}{2\pi^2}k_c \cdot \left(\frac{dB}{dt}\right)^2 \tag{3.49}$$

and

$$p_e(t) = \frac{1}{C_e}k_e \cdot \left|\frac{dB}{dt}\right|^{1.5} \tag{3.50}$$

$C_e = 8.763363$ is from the numerical integration of

$$C_e = (2\pi)^{1.5} \cdot \frac{2}{\pi}\int_0^{\frac{\pi}{2}} \cos^{1.5}\theta d\theta \tag{3.51}$$

For the more general case of $\beta \neq 2$ in (3.41), (3.48) is extended to

$$H_{\text{irr}} = \pm\frac{1}{C_\beta}k_h \cdot |B_m \cos(\theta)|^{\beta-1}, \tag{3.52}$$

where H_{irr} takes the same sign as dB/dt and

$$C_\beta = 4 \int_0^{\frac{\pi}{2}} \cos^\beta \theta d\theta \tag{3.53}$$

As expected, (3.52) becomes (3.48) with $C_\beta = \pi$ when $\beta = 2$.

To obtain B_m from the history record of stored elliptical loops, two rules are applied. One is the wiping-out rule: all ellipses inside the current ellipse are wiped out. The other rule is that if a smaller ellipse (minor loop) is created, the current ellipse is pushed into the recorded ellipse list and the new smaller ellipse is taken as the current ellipse.

In the same way, based on the Steinmetz equation (3.42), the core loss for ferrite material in the time domain is derived as

$$p_v(t) = |K| \cdot \left| \frac{dB}{dt} \right|^\alpha \tag{3.54}$$

where

$$K = \pm \frac{1}{C_{\alpha\beta}} C_m \cdot |B_m \cos(\theta)|^{\beta - \alpha} \tag{3.55}$$

$$C_{\alpha\beta} = (2\pi)^\alpha \cdot \frac{2}{\pi} \int_0^{\frac{\pi}{2}} \cos^\beta \theta d\theta \tag{3.56}$$

When $\alpha = 1$, K has the measure of the magnetic field H, and (3.54) becomes (3.44). In this case, only the hysteresis loss component is considered and K becomes H_{irr}.

In the 3D case, the scalar model (3.44), (3.49), and (3.50) for soft materials is modified as

$$p_h(t) = \left\{ \left| H_x \frac{dB_x}{dt} \right|^{\frac{2}{\beta}} + \left| H_y \frac{dB_y}{dt} \right|^{\frac{2}{\beta}} + \left| H_z \frac{dB_z}{dt} \right|^{\frac{2}{\beta}} \right\}^{\frac{\beta}{2}} \tag{3.57}$$

$$p_c(t) = \frac{1}{2\pi^2} k_c \cdot \left\{ \left(\frac{dB_x}{dt} \right)^2 + \left(\frac{dB_y}{dt} \right)^2 + \left(\frac{dB_z}{dt} \right)^2 \right\} \tag{3.58}$$

$$p_e(t) = \frac{1}{C_e} k_e \cdot \left\{ \left(\frac{dB_x}{dt} \right)^2 + \left(\frac{dB_y}{dt} \right)^2 + \left(\frac{dB_z}{dt} \right)^2 \right\}^{0.75} \tag{3.59}$$

For ferrite materials, the 3D model based on components p_{vx}, $p_{vy,}$ and p_{vz}, from (3.54) are expressed as

$$p_v(t) = \left\{ (p_{vx})^{2/\beta} + (p_{vy})^{2/\beta} + (p_{vz})^{2/\beta} \right\}^{\beta/2} \tag{3.60}$$

Equations (3.57)–(3.60) can also be used for 2D core loss computation when the z component is set to zero.

3.3.2 Core Loss Effects on Magnetic Fields

Basic Field Equation

When the effects of lamination core loss on magnetic field are taken into account, the magnetic field **H** consists of the following components

$$\mathbf{H} = \mathbf{H}_{re} + \mathbf{H}_p = \mathbf{H}_s + \mathbf{T} + \nabla\Omega \tag{3.61}$$

where $\mathbf{H}_s$ corresponds to all exciting current sources, $\mathbf{T}$ is the vector electric potential representing eddy currents in conducting regions, Ω is the scalar magnetic potential [15, 16],

$$\mathbf{H}_{re} = \mathbf{H} - \mathbf{H}_p = [\mu]^{-1}\mathbf{B} \tag{3.62}$$

is the reversible component of the magnetic field associated with normal *B–H* curve without hysteresis loop, the permeability tensor $[\mu]$ is anisotropic in lamination regions (3.25), and

$$\mathbf{H}_p = \mathbf{H}_{pc} + \mathbf{H}_{ph} + \mathbf{H}_{pe} \tag{3.63}$$

is an irreversible additional field component which is introduced to consider the effects of the total core loss on **H** field. Here $\mathbf{H}_{ph}$, $\mathbf{H}_{pc}$, and $\mathbf{H}_{pe}$ are associated with individual effect of static hysteresis loss, classical eddy current loss, and excess loss, respectively.

According to basic field equations

$$\begin{cases} \nabla \times \mathbf{E} = -\dfrac{\partial \mathbf{B}}{\partial t} \\ \nabla \cdot \mathbf{B} = 0 \end{cases} \tag{3.64}$$

and (3.61) and (3.62), we can get the **T**-Ω formulation as [15]

$$\begin{cases} \nabla \times ([\sigma]^{-1}\nabla \times \mathbf{T}) + \dfrac{\partial}{\partial t}[\mu](\mathbf{T} + \nabla\Omega) = -\dfrac{\partial}{\partial t}[\mu](\mathbf{H}_s - \mathbf{H}_p) \\ \nabla \cdot [\mu](\mathbf{T} + \nabla\Omega) = -\nabla \cdot [\mu](\mathbf{H}_s - \mathbf{H}_p) \end{cases}, \tag{3.65}$$

where Ω is the magnetic scalar potential (MSP) in the whole domain and **T** is the electrical vector potential (EVP) in the conducting region.Unlike the source component $\mathbf{H}_s$ that is either known or solved together with voltage and/or coupled circuit equations, the component $\mathbf{H}_p$, consisting of $\mathbf{H}_{pc}$, $\mathbf{H}_{ph}$, and $\mathbf{H}_{pe}$, is unknown. These unknown components can be solved by iteration, or solved together with **T** and Ω.

In solid conductor regions, the conductivity tensor $[\sigma]$in (3.65) is isotropic as given below

$$[\sigma] = \begin{bmatrix} \sigma & 0 & 0 \\ 0 & \sigma & 0 \\ 0 & 0 & \sigma \end{bmatrix} \tag{3.66}$$

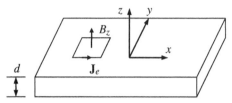

Figure 3.14   Eddy current produced by the normal component of flux density.

Effects of Eddy Current Loss Caused by the Normal Component of Flux Density

When the normal flux density component in the z direction B_z changes, an eddy current field distributed in the x–y plane will be produced, as shown in Figure 3.14.

The equivalent conductivity tensor in lamination regions is anisotropic, and can be expressed as

$$[\sigma] = \begin{bmatrix} k_{\mathrm{lam}}\sigma & 0 & 0 \\ 0 & k_{\mathrm{lam}}\sigma & 0 \\ 0 & 0 & \sigma_{\mathrm{min}} \end{bmatrix} \tag{3.67}$$

where k_{lam} is the lamination factor, and σ_{min}, the minimum conductivity limit, is used to ensure that the system equation is non-singular.

After the lamination eddy current related electric vector potential $\mathbf{T}_e$ is solved in whole lamination region by (3.65), the tangential eddy current $\mathbf{J}_e$ is computed from

$$\mathbf{J}_e = \nabla \times \mathbf{T}_e \tag{3.68}$$

and the loss density is

$$p_e = \frac{\mathbf{J}_e \cdot \mathbf{J}_e}{k_{\mathrm{lam}}\sigma} \tag{3.69}$$

Effects of Eddy Current Loss Caused by the Tangential Components of Flux Density

When the flux density components B_x and/or B_y, tangential to the lamination plane, alternate, the produced eddy current field is bounded inside each lamination, as shown in Figure 3.15.

The effects of eddy current produced by the tangential flux components can be considered by means of an equivalent magnetic field component $\mathbf{H}_{pc}$, which can be computed from the eddy current core loss. From (3.49), the dynamic eddy current core loss per unit volume can be express in the vector form as

$$p_c(t) = [k]\frac{\partial \mathbf{B}}{\partial t} \cdot \frac{\partial \mathbf{B}}{\partial t} \tag{3.70}$$

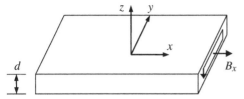

Figure 3.15   Eddy current produced by the tangential components of flux density.

where tensor $[k]$ is defined as

$$[k] = \frac{k_c}{2\pi^2}\begin{bmatrix} 1 & 0 & 0 \\ 0 & 1 & 0 \\ 0 & 0 & 0 \end{bmatrix} \qquad (3.71)$$

On the other hand, the eddy current core loss per unit volume can be expressed as

$$p_c(t) = \mathbf{H}_{pc} \cdot \frac{\partial \mathbf{B}}{\partial t} \qquad (3.72)$$

Therefore, the equivalent magnetic field component $\mathbf{H}_{pc}$ caused by the eddy current loss is

$$\mathbf{H}_{pc} = [k]\frac{\partial \mathbf{B}}{\partial t} \qquad (3.73)$$

Effects of Hysteresis Loss
In the time domain, the hysteresis core loss can be calculated based on the equivalent magnetic field component $\mathbf{H}_{ph}$, as

$$p_h(t) = \mathbf{H}_{ph} \cdot \frac{\partial \mathbf{B}}{\partial t} \qquad (3.74)$$

Each coordinate component of $\mathbf{H}_{ph}$ can be independently computed by the EEL as

$$H_{ph} = \pm H_m\sqrt{1 - (B/B_m)^2} = \pm\frac{k_h}{\pi}\sqrt{B_m^2 - B^2}, \qquad (3.75)$$

where "+" and "−" correspond to the ascending and descending branch, respectively.

Effects of Excess Loss
From (3.48), the excess core loss can be expressed in the time domain in the vector form as

$$p_e(t) = \frac{k_e}{C_e}\left[\left(\frac{\partial B}{\partial t}\right)_m\right]^{-0.5}\frac{\partial \mathbf{B}}{\partial t} \cdot \frac{\partial \mathbf{B}}{\partial t}, \qquad (3.76)$$

where C_e is obtained from (3.51), and

$$\left(\frac{\partial B}{\partial t}\right)_m = \sqrt{\left(\frac{\partial B_x}{\partial t}\right)^2 + \left(\frac{\partial B_y}{\partial t}\right)^2 + \left(\frac{\partial B_z}{\partial t}\right)^2} \tag{3.77}$$

Comparing (3.76) with

$$p_e(t) = \mathbf{H}_{pe} \cdot \frac{\partial \mathbf{B}}{\partial t} \tag{3.78}$$

we get

$$\mathbf{H}_{pe} = \frac{k_e}{C_e}\left[\left(\frac{\partial B}{\partial t}\right)_m\right]^{-0.5} \frac{\partial \mathbf{B}}{\partial t} \tag{3.79}$$

3.3.3   Implementation of Core Loss Effects

Eddy Current Loss Effects

The discrete format of (3.73) is

$$\mathbf{H}_{pc} = [k]\frac{\mathbf{B} - \mathbf{B}_0}{\Delta t}, \tag{3.80}$$

where $\mathbf{B}_0$ is the field solution of the previous time step.

From (3.61)–(3.63) and (3.80), we have

$$\begin{aligned} \mathbf{H} &= [\mu]^{-1}\mathbf{B} + \left([k]\frac{\mathbf{B} - \mathbf{B}_0}{\Delta t} + \mathbf{H}_{ph} + \mathbf{H}_{pe}\right), \\ &= [\mu_{eq}]^{-1}\mathbf{B} + \mathbf{H}'_p \end{aligned} \tag{3.81}$$

where

$$[\mu_{eq}] = \begin{bmatrix} \mu_x/k_{\mu x} & 0 & 0 \\ 0 & \mu_y/k_{\mu y} & 0 \\ 0 & 0 & \mu_z \end{bmatrix} \tag{3.82}$$

is the equivalent permeability tensor, and

$$\mathbf{H}'_p = \mathbf{H}_{pc0} + \mathbf{H}_{ph} + \mathbf{H}_{pe} \tag{3.83}$$

with

$$\begin{cases} k_{\mu x} = 1 + \dfrac{k_c \mu_x}{2\pi^2 \Delta t} \\[2ex] k_{\mu y} = 1 + \dfrac{k_c \mu_y}{2\pi^2 \Delta t} \end{cases} \tag{3.84}$$

and

$$\mathbf{H}_{pc0} = -[k]\mathbf{B}_0/\Delta t \tag{3.85}$$

With the introduction of the equivalent permeability, (3.65) can be rewritten as

$$
\begin{cases}
\nabla \times ([\sigma]^{-1} \nabla \times \mathbf{T}) + \dfrac{\partial}{\partial t}[\mu_{eq}](\mathbf{T} + \nabla\Omega) = -\dfrac{\partial}{\partial t}[\mu_{eq}](\mathbf{H}_s - \mathbf{H}'_p) \\[2mm]
\nabla \cdot [\mu_{eq}](\mathbf{T} + \nabla\Omega) = -\nabla \cdot [\mu_{eq}](\mathbf{H}_s - \mathbf{H}'_p)
\end{cases}
\tag{3.86}
$$

After (3.86) has been solved, $\mathbf{H}$ and $\mathbf{B}$ can be further obtained by

$$
\begin{cases}
\mathbf{H} = \mathbf{H}_s + \mathbf{T} + \nabla\Omega \\[2mm]
\mathbf{B} = [\mu_{eq}](\mathbf{H} - \mathbf{H}'_p)
\end{cases}
\tag{3.87}
$$

Hysteresis and Excess Loss Effects

The impact of the eddy current loss component on field solutions has been directly incorporated into the original field equations in terms of the equivalent permeability $[\mu_{eq}]$ and the modified eddy current loss field component $\mathbf{H}_{pc0}$ without involving an iteration process. To further take into account the effects of the other two components: hysteresis loss component in terms of $\mathbf{H}_{ph}$ and the excess loss component in terms of $\mathbf{H}_{pe}$, an iterative procedure can be applied.

If the flux density at time t_0 is $\mathbf{B}_0$, and the corresponding field components due to hysteresis and excess core losses are $\mathbf{H}_{ph0}$ and $\mathbf{H}_{pe0}$, respectively (the starting values for $\mathbf{H}_{ph0}$ and $\mathbf{H}_{pe0}$ are 0 because the relevant losses at $t = 0$ are 0), then the total equivalent irreversible field component at time t_0 is

$$
\mathbf{H}'_{p0} = \mathbf{H}_{pc0} + \mathbf{H}_{ph0} + \mathbf{H}_{pe0}
\tag{3.88}
$$

and the flux density $\mathbf{B}_1$ and field component $\mathbf{H}'_{p1}$

$$
\mathbf{H}'_{p1} = \mathbf{H}_{pc0} + \mathbf{H}_{ph1} + \mathbf{H}_{pe1}
\tag{3.89}
$$

at time t_1 with time step of $\Delta t = (t_1 - t_0)$ can be solved through the following process:

1. Let $\mathbf{H}'_p = \mathbf{H}'_{p0}$, which is obtained from (3.88), solve the nonlinear equations of (3.86) based on Newton–Raphson iteration, get flux density $\mathbf{B}_1$ by (3.87);

2. Freeze material properties $[\mu_{eq}]$ at the solved field point $\mathbf{B}_1$, that is, linearize all nonlinear materials using $\mu = B_1/H_1$;

3. Get $\mathbf{H}'_{p1}$ from $\mathbf{B}_1$ based on (3.89);

4. If the error between $\mathbf{H}'_p$ and $\mathbf{H}'_{p1}$ is not acceptable, let $\mathbf{H}'_p = \mathbf{H}'_{p1}$, solve linear equation of (3.86) based on the frozen material parameters, get flux density $\mathbf{B}_1$, go back to 3;

5. If the error between $\mathbf{H}'_p$ and $\mathbf{H}'_{p1}$ is acceptable, stop the process.

The above iteration process can be described by the flowchart as shown in Figure 3.16. Figure 3.16 shows that in every time step, there are two iterations: one is the Newton–Raphson iteration which solves material nonlinearity at $\mathbf{H}'_{p0}$ with ignoring the saturation change due to $\mathbf{H}'_{p1} - \mathbf{H}'_{p0}$; the other iteration solves the field component $\mathbf{H}'_{p1}$ with nonlinear materials frozen at the saturation with $\mathbf{H}_{p0}$. This means

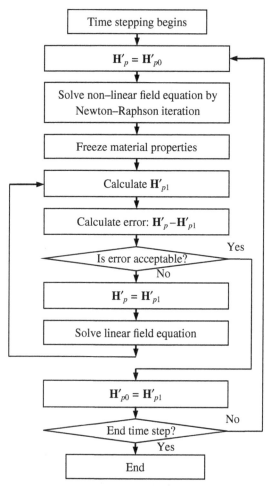

Figure 3.16 Iterative process considering the lamination core loss effects.

that the saturation change due to $\mathbf{H}'_p$ change has one time-step lagging ensuring that the second iteration does not affect the convergence of the Newton–Raphson iteration. For a reasonable time-step size, the saturation change due to $\mathbf{H}'_p$ change in one time step is not sensitive to the time-step interval.

3.3.4 Case Study: Electrical and Mechanical Power-Balance Tests

According to the power-balance principle, when core loss effects are taken into account, an additional input power is required to balance the core loss. Thus, the presented approach can be validated by power-balance testing: the increase of input power (electrical power and/or mechanical power) between with and without considering the core loss effects should be equal to the total core loss. Since the objective

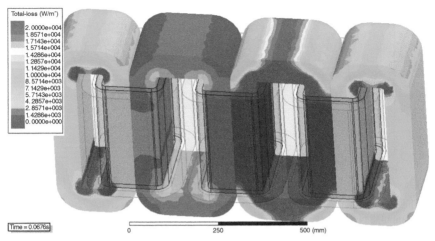

Figure 3.17   Instantaneous core loss distribution of the 250 kVA three-phase amorphous metal power transformer.

here is to validate the effectiveness of taking into account core loss effects, which is independent of load conditions, it is more convenient and accurate to do the validation at no-load condition. Under load conditions, it is unavoidable that the core loss effects will somehow cause output power change even though this change may be not significant.

Two application examples are presented for the power-balance testing. The first application is for the computation of core loss of 250 kVA three-phase amorphous metal power transformer with five legs. The delta-connected three-phase primary windings are energized by three-phase voltage sources, and the secondary windings are open circuit. The instantaneous core loss distribution is shown in Figure 3.17.

The primary phase currents are computed under the following two cases: with and without considering the core loss effects. The increase of phase currents due to core loss effects can be derived by subtracting the phase currents without considering core loss effects from those considering core loss effects. The input power increase due to considering core loss effects is obtained from

$$\Delta p_{in} = \Delta i_A \cdot e_A + \Delta i_B \cdot e_B + \Delta i_C \cdot e_C, \tag{3.90}$$

where Δi_A, Δi_B, and Δi_C represent the increases of three phase currents, and e_A, e_B, and e_C are the three phase induced voltages. The input power increase derived from (3.90) is compared with the computed core loss in Figure 3.18. The average input power increase and the computed core loss in the last period (80–100 ms) are compared with the measured data in Table 3.2.

The second application is for the no-load core loss computation of a 165 W, 4-pole interior permanent magnet (IPM) brushless DC (BLDC) motor. At the ideal no-load operation, the stator currents are 0. To simplify the 3D model, the stator windings

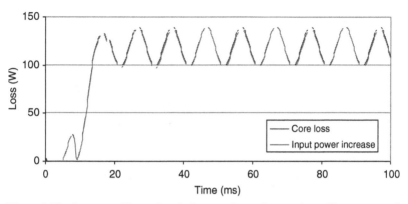

Figure 3.18 Increase of input electrical power due to the core loss effects compared with the computed core loss ($k_h = 21.08$, $k_c = 0$, $k_e = 0$).

TABLE 3.2 Computed and measured core loss of the 250 kVA three-phase amorphous metal power transformer

	Value	Unit
Input power increase	118	W
Computed core loss	119	W
Measured core loss	126	W

are removed from the stator core. According to the periodic condition, only one pole (90°) is required for FEA. The meshed 3D model is shown in Figure 3.19.

At the no-load operation without considering the core loss effects, the electromagnetic torque of the machine has only the cogging torque component. The average

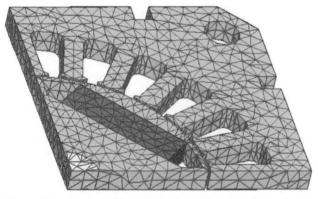

Figure 3.19 Meshed 3D model (without the stator windings) of the 165 W, 4-pole BLDC motor.

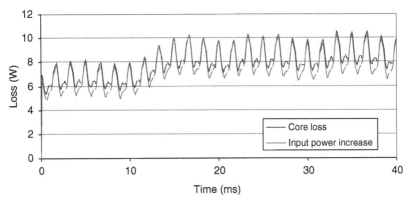

Figure 3.20 Input mechanical power increase due to the core loss effects compared with the computed core loss ($k_h = 260.4$, $k_c = 0.822$, $k_e = 40.54$).

value of the cogging torque is zero. When the core loss effects are taken into account, there exists an additional torque component, or torque increase, due to the core loss. The waveform of the torque increase can be derived by subtracting the torque waveform without the core loss effects from that with core loss effects.

The mechanical input power increase, derived from the constant rotor speed and the torque increase due to the core loss effects, is compared with the computed core loss in Figure 3.20. The average input power increase and the computed core loss in the last period (20–40 ms) are compared with the measured data in Table 3.3.

3.4 VECTOR HYSTERESIS MODEL

Hysteresis loss computed from (3.41) is valid only when the magnetic field is not deeply saturated. At deeply saturated field, even though the flux density B will increase linearly with the applied field intensity H, with the slope of the free-space permeability μ_0, the magnetization M will keep constant. In such cases, for alternating magnetic fields, the hysteresis loss will not continue to increase as the magnitude of the alternating flux density increases. For rotating magnetic fields, the hysteresis loss will drop down to zero when the field is fully saturated. This property is called "the rotational loss property" [17–19].

TABLE 3.3 Computed and measured core loss of the 165 W 4-pole IPM BLDC motor

	Value	Unit
Input power increase	8.1	W
Computed core loss	8.5	W
Measured core loss	8.7	W

To predict the magnetization behavior for isotropic magnetic materials with hysteresis in 2D or 3D transient FEA, it has been recognized that the vector play model [20–23] is more computationally efficient than various vector Preisach models [17, 24–26]. However, the ordinary vector play model does not obey the rotational loss property.

The ordinary vector play model will be introduced first in this section. Then an improved vector play model [27] is presented to predict the magnetization behavior for isotropic magnetic materials with hysteresis. All required parameters of the model can be directly identified from the major hysteresis loop. This model not only satisfies the rotational loss property, but also improves the accuracy of the core loss computation at alternating fields.

3.4.1 Ordinary Vector Play Model

The output of a scalar play model is given by

$$m(t) = \sum_{k=1}^{n} f_k(h_{rek}(t)), \tag{3.91}$$

where

$$h_{rek}(t) = P_{\sigma_k}[h(t)] \tag{3.92}$$

is the play operator, the input $h(t)$ is the field density, and $f_k(\cdot)$'s are usual anhysteretic nonlinear functions. The play model (3.91) can be derived from the scalar Preisach model [28].

The simplest scalar play model is with only one play operator. If the anhysteretic nonlinear function $f_k(\cdot)$ is expressed as $M_{an}(\cdot)$, (3.91) is simplified as

$$m = M_{an}(h_{re}), \tag{3.93}$$

where

$$h_{re} = P_r[h] = \max(\min(h_{re0}, h+r), h-r) \tag{3.94}$$

or alternately

$$h_{re} = h - \frac{r(h - h_{re0})}{\max(r, |h - h_{re0}|)} \tag{3.95}$$

with h_{re0} being the initial value of h_{re}, and r, representing the intrinsic coercivity σ_k in (3.92), is a pre-determined parameter. The scalar operator can be illustrated by Figure 3.21.

The basic idea of the play model is to decompose the applied field intensity h into two components, the reversible component h_{re}, and the irreversible components h_{ir}. The magnetization m varies with the reversible field component along an anhysteretic curve, and the trace of the magnetization alternately varying with the resultant field intensity in a fixed direction forms a hysteresis loop. The key task of the play model is to decompose the field intensity and return the reversible component. After

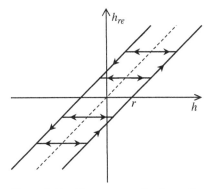

Figure 3.21   Play operator for the ordinary play model.

the reversible field component is obtained from the play operator (3.95), the magnetization m can be obtained from a pre-determined anhysteretic curve (3.93), as shown in Figure 3.22.

The play operator expressed in (3.95) can be extended for vector field as

$$\mathbf{h}_{re} = \mathbf{h} - \frac{r(\mathbf{h} - \mathbf{h}_{re0})}{\max(r, |\mathbf{h} - \mathbf{h}_{re0}|)} \qquad (3.96)$$

or alternately expressed as

$$\mathbf{h}_{re} = \begin{cases} \mathbf{h}_{re0} & \text{if } |\mathbf{h} - \mathbf{h}_{re0}| < r \\ \mathbf{h} - r \cdot \dfrac{\mathbf{h} - \mathbf{h}_{re0}}{|\mathbf{h} - \mathbf{h}_{re0}|} & \text{if } |\mathbf{h} - \mathbf{h}_{re0}| \geq r \end{cases} \qquad (3.97)$$

The vector magnetization $\mathbf{m}$ is

$$\mathbf{m} = M_{an}(h_{re}) \cdot \mathbf{h}_{re}/h_{re} \qquad (3.98)$$

where h_{re} is the absolute value of $\mathbf{h}_{re}$.

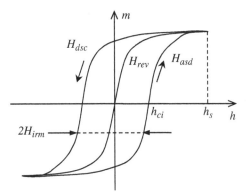

Figure 3.22   Magnetizing curves (H_{rev}, H_{asd}, and H_{dsc} represent anhysteretic, ascending, and descending curves, respectively).

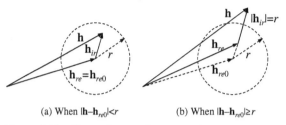

(a) When $|h-h_{re0}|<r$ (b) When $|h-h_{re0}|\geq r$

Figure 3.23   Vector diagram for vector play operator.

The vector play operator of (3.97) can be illustrated by the vector diagram as shown in Figure 3.23.

Draw a circle at the tip of vector $\mathbf{h}_{re0}$ with radius r. If the tip of the applied field $\mathbf{h}$ falls inside the circle, keep the reversible component unchanged, as shown in Figure 3.23a; otherwise, get $\mathbf{h}_{ir}$ in the direction of $\mathbf{h} - \mathbf{h}_{re0}$ with length of r, and then let $\mathbf{h}_{re} = \mathbf{h} - \mathbf{h}_{ir}$, as shown in Figure 3.23b.

3.4.2   Improved Vector Play Model

Play Operator

In Figure 3.23, if the applied field rotates, it can be proved that at the steady state, the irreversible component $\mathbf{h}_{ir}$ will be perpendicular to the reversible component $\mathbf{h}_{re}$. In the ordinary model, the magnitude of $\mathbf{h}_{ir}$ is constant no matter how large the applied field is, which means $\mathbf{m}$, in the same direction of $\mathbf{h}_{re}$, will always lag $\mathbf{h}$ a certain angle. Therefore, the ordinary vector play model does not satisfy the rotational loss property.

The model can be modified to satisfy the rotational loss property by defining r as a function of the reversible field component h_{re} with $r = 0$ when $h_{re} \geq h_s$, here h_s is the saturation field. The vector play operator then becomes

$$\mathbf{h}_{re} = \begin{cases} \mathbf{h}_{re0} & \text{if } |\mathbf{h} - \mathbf{h}_{re0}| < r(h_{re0}) \\ \mathbf{h} - r(h_{re}) \cdot \dfrac{\mathbf{h} - \mathbf{h}_{re0}}{|\mathbf{h} - \mathbf{h}_{re0}|} & \text{if } |\mathbf{h} - \mathbf{h}_{re0}| \geq r(h_{re0}) \end{cases} \tag{3.99}$$

The play operator for the improved vector play model is shown in Figure 3.24.

The parameters for the improved vector play model, including $M_{an}(h_{re})$ and $r(h_{re})$, are identified from the major hysteresis loop. The major hysteresis loop consists of the ascending branch $M_{asd}(h)$ and the descending branch $M_{dec}(h)$. The ascending, or descending, curve can be directly obtained from each other based on the odd symmetry condition, and therefore, only one branch is required from input.

If the inverse functions of $M_{asd}(h)$ and $M_{dsc}(h)$ are denoted as $H_{asd}(m)$ and $H_{dsc}(m)$, respectively, as shown in Figure 3.22, then

$$\begin{cases} H_{rev}(m) = [H_{asd}(m) + H_{dsc}(m)]/2 \\ H_{irm}(m) = [H_{asd}(m) - H_{dsc}(m)]/2 \end{cases} \tag{3.100}$$

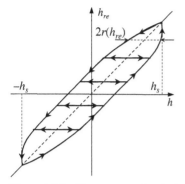

Figure 3.24   Play operator for the improved play model.

From (3.100), we finally obtain

$$\begin{cases} M_{an}(h_{re}) = H_{rev}{}^{-1}(h_{re}) \\ r(h_{re}) = H_{irm}(M_{an}(h_{re})) \end{cases} \tag{3.101}$$

Local Iteration

In the vector play model, the flux density **b** at the applied field **h** is expressed as

$$\mathbf{b} = \mu_0(\mathbf{m} + \mathbf{h}), \tag{3.102}$$

where **m**, defined by (3.98), is a function of $\mathbf{h}_{re}$ which is obtained by solving (3.99). Since r depends on h_{re}, a local iterating process is required to solve (3.99).

We can derive **b** from **h** based on the following local iteration algorithm. When the applied field **h** locates outside the circle as shown in Figure 3.23b, we can derive $\mathbf{h}_{re}$ according to the following iteration process:

1. Assume $\mathbf{h}_{re} = \mathbf{h}_{re0}$;
2. Get $\mathbf{h}_{ir}$ from $r(h_{re})$ curve and the direction of $(\mathbf{h} - \mathbf{h}_{re})$;
3. Calculate $\Delta \mathbf{h} = \mathbf{h} - (\mathbf{h}_{re} + \mathbf{h}_{ir})$;
4. Let $\mathbf{h}_{re} = \mathbf{h}_{re} + \alpha \, \Delta \mathbf{h}$;
5. Repeat steps 2~4 until $|\Delta \mathbf{h}|/h_s < \varepsilon$.

In the above iterating process, α is a relaxation factor which can be optimized based on the historic iterating results, and ε is the given tolerance. After $\mathbf{h}_{re}$ is obtained, **m** and **b** are computed from (3.98) and (3.102), respectively.

We can also derive **h** from **b**. From (3.98) and (3.102), $\mathbf{b}_{re0}$ can be obtained from $\mathbf{h}_{re0}$. Let us draw a circle at the tip of vector $\mathbf{b}_{re0}$ with the radius of $r_b = \mu_0 r(h_{re0})$, as shown in Figure 3.25. When the tip of vector **b** locates inside the circle, let $\mathbf{h}_{re} = \mathbf{h}_{re0}$, otherwise, follow the iterating process as listed below:

1. Assume $\mathbf{b}_{re} = \mathbf{b}$;
2. Get $\mathbf{h}_{re}$ from $\mathbf{b}_{re}$ based on the anhysteretic B–H curve;

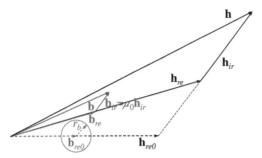

Figure 3.25   Vector diagram to derive **h** from **b**.

3. Get $\mathbf{h}_{ir}$ from $r(h_{re})$ curve and the direction of $(\mathbf{h}_{re} - \mathbf{h}_{re0})$;

4. Let $\mathbf{b}_{ir} = \mu_0\,\mathbf{h}_{ir}$;

5. Calculate $\Delta\mathbf{b} = \mathbf{b} - (\mathbf{b}_{re} + \mathbf{b}_{ir})$;

6. Let $\mathbf{b}_{re} = \mathbf{b}_{re} + \alpha\,\Delta\mathbf{b}$;

7. Repeat steps 2~6 until $|\Delta\mathbf{b}|/b_s < \varepsilon$.

In step 7, b_s is the flux density at h_s. After $\mathbf{h}_{re}$ is obtained, $\mathbf{h}$ is computed from

$$\mathbf{h} = \mathbf{h}_{re} + (\mathbf{b} - \mathbf{b}_{re})/\mu_0 \qquad (3.103)$$

Optimal Relaxation Factor

Consider a nonlinear vector equation

$$\mathbf{v} = \mathbf{f}(\mathbf{v}) \qquad (3.104)$$

The iteration algorithm to solve (3.104) can be expressed as

$$\mathbf{v}_i = \mathbf{f}(\mathbf{v}_i') \qquad i = 0, 1, \ldots, \qquad (3.105)$$

where $\mathbf{v}_i'$ and $\mathbf{v}_i$ are the assumed and estimated values of **v** at step i, respectively. The assumed value for next iteration can be predicted by

$$\mathbf{v}_{i+1}' = \mathbf{v}_i' + \alpha\,\Delta\mathbf{v}_i \qquad (3.106)$$

or

$$\mathbf{v}_{i+1}' = \mathbf{v}_{i-1}' + \alpha\,\Delta\mathbf{v}_{i-1} \qquad (3.107)$$

where α is a relaxation factor to be determined, and

$$\Delta\mathbf{v}_k = \mathbf{v}_k - \mathbf{v}_k' \qquad k = i, i-1. \qquad (3.108)$$

Minimizing the difference between (3.106) and (3.107), that is

$$\varepsilon_v = \left|\left(\mathbf{v}_i' + \alpha\,\Delta\mathbf{v}_i\right) - \left(\mathbf{v}_{i-1}' + \alpha\,\Delta\mathbf{v}_{i-1}\right)\right|^2 = \min \qquad (3.109)$$

TABLE 3.4   Comparison between the average number of iterations using and without using the optimal relaxation factors

	With $\alpha = 1$	With optimal α
Derive **b** from **h**	33.2	4.8
Derive **h** from **b**	7.5	3.2

we obtain the optimal relaxation factor as

$$\alpha = -\frac{\left(\mathbf{v}'_i - \mathbf{v}'_{i-1}\right) \cdot (\Delta\mathbf{v}_i - \Delta\mathbf{v}_{i-1})}{(\Delta\mathbf{v}_i - \Delta\mathbf{v}_{i-1}) \cdot (\Delta\mathbf{v}_i - \Delta\mathbf{v}_{i-1})} \tag{3.110}$$

It requires the results of previous two iterations to calculate α from (3.110). After the first iteration with $i = 0$, (3.110) is invalid, and α can be assumed to be 1. Table 3.4 compares the average number of local iterations, using and without using the optimal relaxation factor, in a typical field solution with $\varepsilon = 10^{-12}$.

Using the optimal relaxation factors significantly speeds up the convergence of the local iteration, especially in getting **b** from **h**. Extra attention must be paid to the convergence of the local iteration because it is essential for the convergence in solving the nonlinear field equations.

Validations

A measured major loop cited from [29] is used as input. The simulated major loop using the improved model is shown in Figure 3.26, compared with the measured data. Since the parameters are identified directly from the measured major loop, the simulated results are identical with the measured data. The major-loop energy loss per cycle simulated by the improved play model, together with that by the ordinary play model, is compared with the measured data in Table 3.5. The steady-state rotational

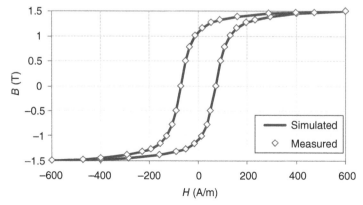

Figure 3.26   Simulated major loop by the improved play model compared with the measured one.

TABLE 3.5   Major-loop energy losses at alternating fields

	Energy loss (J/m^3)
Simulated by ordinary play model	426.7
Simulated by improved play model	416.5
Measured	416.5

loss simulated by the improved play model is compared with that by the ordinary play model in Figure 3.27.

3.4.3   Adaptive Fixed Point Iteration Algorithm

To solve nonlinear magnetic field problems, the employment of iteration is inevitable. Although the Newton–Raphson (NR) method with fast quadratic rate has been widely used to solve a nonlinear system, it often suffers from instability because it is very sensitive to the derivative at the temporary solution during the iteration.

For a time-stepping transient magnetic field problem with hysteresis media involved, the field solution depends on the historic conditions. When the last field solution of the previous time step is completed, the derivative dB/dH at the last field solution is not well defined because it may have two significantly different values in two field changing (increasing and decreasing) directions. Therefore, the NR method is not directly applicable to hysteric problems. Instead, the fixed point method plays an important role in the iteration of nonlinear problems with hysteresis [30–34].

There are two schemes in the fixed point method for magnetic field computation, the B-correction and H-correction schemes. The former uses curve $H(B)$, and the latter uses $B(H)$, during the iteration. In this section, the fixed point method with B-correction and H-correction schemes is reviewed first. Then an adaptive fixed point iteration algorithm by alternately using the B-correction and H-correction schemes is introduced.

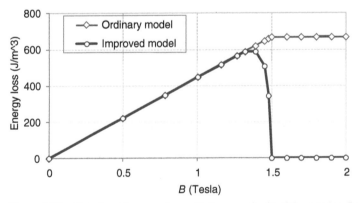

Figure 3.27   Rotational loss varying with the magnitude of the rotating flux density.

Fixed Point Method

For any continuous nonlinear function $y = f(x)$, if y is known as y_0, the equation can be rewritten as $x = F(x)$. The root of the equation can be computed by the fixed point iteration as

$$x_{k+1} = F(x_k), \quad k = 0, 1, 2, \ldots \tag{3.111}$$

For any period $[a, b]$, if

$$|F(b) - F(a)| \leq L|b - a| \tag{3.112}$$

the iteration of (3.111) will converge to a fixed point, as long as $L < 1$, because

$$|x_{k+1} - x_k| = |F(x_k) - F(x_{k-1})| \leq L|x_k - x_{k-1}|$$
$$\leq L^k|x_1 - x_0| \tag{3.113}$$

Equation (3.113) shows that the smaller the L is, the faster the iteration will converge. Function $F(x)$ can be constructed in many schemes. One possible scheme is

$$F(x) = x + \frac{y_0 - f(x)}{c}, \tag{3.114}$$

where c can be any constant provided (3.112) is satisfied. It can be clearly seen from (3.114) that as long as $F(x)$ is converged to x, the original equation $y_0 = f(x)$ is satisfied.

To better appreciate the fixed point iteration algorithm, let us take an inductor with uniform cross-section core excited by a coil of N turns carrying a current of $i(t)$ as an illustration example. If the core is treated as a one-dimensional (1D) element and the magnetic property is expressed as $H(B)$, the fixed point iteration in the B-correction scheme is

$$B_{k+1} = B_k + \frac{H_a - H(B_k)}{\nu_{FP}} = F(B_k), \tag{3.115}$$

where $H_a = Ni(t)/l$ is constant during the iteration with l being the average length of the core. The constant reluctivity ν_{FP} can freely be selected provided that (3.112) is satisfied. The global-coefficient method [30] suggested

$$\nu_{FP} = \frac{\nu_{d\,max} + \nu_{d\,min}}{2} \approx \frac{1}{2\mu_0}, \tag{3.116}$$

where ν_{dmax} and ν_{dmin} are the maximum and minimum differential reluctivities of the curve $H(B)$, respectively. The slope L of curve $F(B)$ is smaller than, but close to, 1.0 for whole region, see Figure 3.28b. Therefore, the iteration using (3.116) will converge at any starting point. However, it suffers very slow convergence no matter the fixed point is in the unsaturated or in the saturated region.

If the constant reluctivity is selected as

$$\nu_{FP} = \nu_{d\,max} = \frac{1}{\mu_0} \tag{3.117}$$

the iteration will converge very fast when the fixed point is in the saturated region where the slope L is close to 0, see Figure 3.29a. When the fixed point is in the

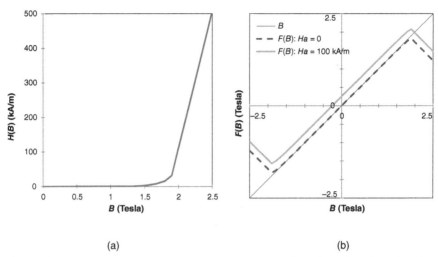

(a)   (b)

Figure 3.28   Iterative function: (a) $H(B)$ curve; (b) B-correction scheme based on (3.115) and (3.116).

unsaturated region, convergence is also guaranteed, but with slow convergence rate because L is close to 1.0.

When magnetic property is expressed as $B(H)$, the fixed point iteration using H-correction scheme is expressed as

$$H_{k+1} = H_k + \frac{B_a - B(H_k)}{\mu_{FP}} = F(H_k), \tag{3.118}$$

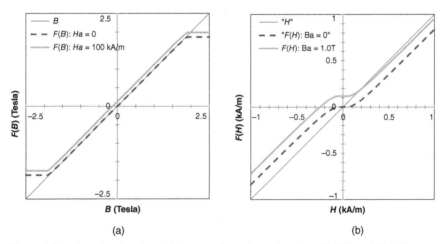

(a)   (b)

Figure 3.29   Iterative function: (a) B-correction scheme based on (3.115) and (3.117); (b) H-correction scheme based on (3.118) and (3.119).

where B_a is the applied flux density which is constant during the iteration. The constant permeability μ_{FP} can be selected as

$$\mu_{FP} = \mu_{d\,max}, \qquad (3.119)$$

where $\mu_{d\text{max}}$ is the maximum differential permeability of the curve $B(H)$. In such a case, the convergence is guaranteed. However, the performance using the H-correction scheme is different in the saturated region with fast convergence rate and in the unsaturated region with very slow convergence rate, as indicated in Figure 3.29b.

In practical applications of 2D/3D FEA, some numerical noise, such as meshing discretization error is inevitable. When the slope L of curve $F(B)$, or $F(H)$, is close to 1.0, such noise may be enlarged in the iteration process due to the cross effects of neighboring mesh elements. As a result, the convergence may no longer be guaranteed. Therefore, to ensure a stable convergence, it is desirable to have the slope L as close to 0 as possible, such as in the saturated region of Figure 3.29a, and in the non-saturated region with small field quantity of Figure 3.29b.

Adaptive Fixed Point Algorithm

As discussed above, the convergence behavior using different schemes is different in different regions. In fact, if the coil in the above-discussed inductor example is excited by a current source, the applied field intensity H_a will be constant in each time step, and thus iteration is not required if the H-correction scheme is used. On the other hand, if the coil is excited by a voltage source, the flux linkage, thus the applied flux density B_a, can be considered as given with the assumption that the impact of coil resistance is trivial. In such a case, the iteration process is also not required if the B-correction scheme is used.

From the above observation, we propose an adaptive fixed point iteration algorithm to alternately use the B-correction scheme and H-correction scheme so as to speed up convergence and improve the stability. The iteration may start with the B-correction scheme in which the constant reluctivity is set to the maximum differential reluctivity, or the minimum differential permeability μ_0. If the solution is not converged to a given accuracy after a certain preset number of iterations, the iteration will be continued by switching to the H-correction scheme in which the constant permeability is set to the maximum differential permeability. With the combined use of the two correction schemes during the entire iteration process, the solution with the minimum error together with the scheme type will be recorded and used as the final solution at the current time step. At the same time, the recorded scheme type will be used as the initial scheme type for the next time step. The flowchart of the iteration process is shown in Figure 3.30.

Clearly, it is a prerequisite that in order to utilize the adaptive fixed point iteration algorithm, the magnetic property must be expressible in both forms of $\mathbf{H}(\mathbf{B})$ and $\mathbf{B}(\mathbf{H})$. For the improved vector play model, a dedicated and very efficient local iteration scheme is developed to satisfy this request (see Section 3.4.2) [27].

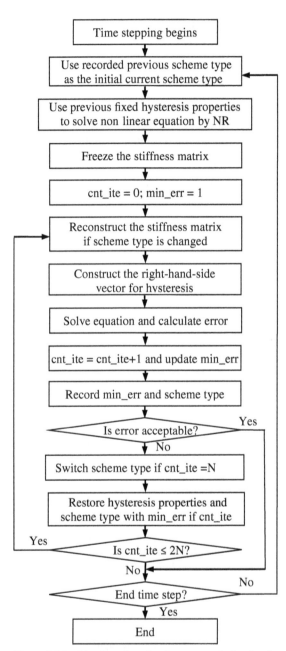

Figure 3.30   The flowchart for the proposed adaptive fixed point iteration (NR is the Newton–Raphson iteration [35]).

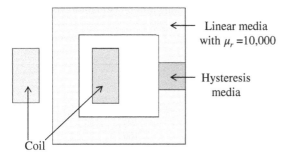

Linear media
with μ_r =10,000

Hysteresis
media

Coil

Figure 3.31   The 2D geometry of the inductor with hysteresis core.

Validation

To validate the effectiveness of the adaptive algorithm, an inductor as shown in Figure 3.31 is taken as a benchmark to compare the average iteration number among different iteration types. The input major hysteresis loop and the simulated loop by the vector hysteresis model under a voltage source excitation are shown in Figure 3.32. Table 3.6 shows the average iteration number using the adaptive algorithm compared with those using either *B*- or *H*-correction scheme under sinusoidal current and voltage excitations, respectively. The maximum iteration number is set to 1000, and the error target is set to 10^{-4}.

Table 3.6 shows that the *B*-correction scheme converges for all time steps when the inductor is excited by a voltage source, but does not converge for some time steps with current excitation. On the other hand, the *H*-correction scheme converges for all time steps when the inductor is excited by a current source, but does not converge for some time steps with voltage excitation. The adaptive algorithm with the combined

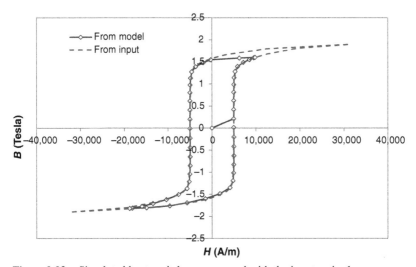

H (A/m)

Figure 3.32   Simulated hysteresis loop compared with the input major loop.

TABLE 3.6   Comparison of average iteration number among different iteration types

Iteration type	Source type	Average iteration number	Total time steps	Non-converged time steps
B-correction	Current	96.47	100	6
scheme	Voltage	23.23	100	0
H-correction	Current	37.77	100	0
scheme	Voltage	124.82	100	5
Adaptive	Current	37.58	100	0
algorithm	Voltage	23.23	100	0

use of both schemes converges for all time steps no matter it is under current or voltage excitations.

3.5   DEMAGNETIZATION OF PERMANENT MAGNETS

Permanent magnet (PM) machines are usually designed with reasonable demagnetization in normal working conditions. In some occasional cases, a PM motor may operate at an overloaded torque. In such a case, the motor may be a little more demagnetized at the very beginning, which will cause the motor start to draw more current to produce the required torque. The increased current may cause even more demagnetization, which will again increase the current. Long-time overload may cause the motor to get overheated. Rising in temperature may drop the intrinsic coercivity and the residual of the permanent magnets, and make them more prone to demagnetization [36, 37]. Therefore, a suitable model is required to study whether this process keeps on going until the machine stalls, or the demagnetization stops at some balanced point causing degraded machine performance.

In this section, after introducing an efficient searching algorithm to iteratively identify a new demagnetization point in a linear model that handles the complete demagnetization curve, we extend the model to take into account the temperature dependence of demagnetization behaviors [38]. The new demagnetization point, or the new worst working point, is searched in each time step during the entire transient solution process. Finally, we present an algorithm to construct demagnetization curves for magnetization.

3.5.1   Demagnetization Curve

When a slowly changed alternating magnetic force is applied to a virgin permanent magnet (PM) material in a fixed direction, the instant magnetic field will trace a hysteresis loop. The curve in the second (or fourth) quadrant of a hysteresis loop is termed demagnetization curve. There are four most important parameters to represent the magnetic characteristics of permanent magnets. The two extreme positions on the demagnetization curve are the two significant parameters. The value of the magnetic flux density corresponding to zero magnetic field intensity is termed

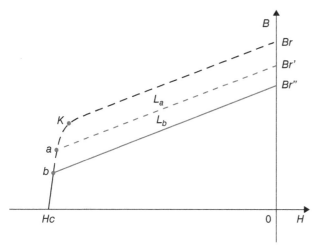

Figure 3.33   Irreversible demagnetization due to working point below knee point.

residual flux density (or remanence induction) B_r; the value of the magnetic field intensity corresponding to zero magnetic flux density is termed coercivity (or coercive force) H_c. The product of the magnetic flux density and the magnetic field intensity at any point on the demagnetization curve is termed magnetic energy product (BH). At the two extreme positions $(0, B_r)$ and $(H_c, 0)$, the magnetic energy product is equal to zero. Somewhere at an intermediate position, the magnetic energy reaches its maximum value and is termed maximum magnetic energy product $(BH)_{\mathrm{max}}$, which is another important parameter. The fourth important parameter is called relative recoil permeability μ_r, which is assumed to be equal to the slope of the tangent line of the demagnetization curve at point $(0, B_r)$.

The initial working point is assumed to be under the condition that the PM is just magnetized and the external magnetization field is removed. When an external demagnetization field is applied, the working point will move along the demagnetization curve toward $(H_c, 0)$ from its initial working point. If the new working point a goes below the knee point K by the external demagnetization field H_a, even after the external field H_a is reduced or totally removed, the subsequent working point will no longer lie along the original B–H curve, but be along a recoil line L_a. As long as the subsequent applied external demagnetization field intensity does not exceed H_a, the permanent magnet will work along the recoil line L_a. If, however, a greater external demagnetization field H_b is applied, a lower demagnetization point b associated with a new recoil line L_b will be established. This suggests that the portion of the original demagnetization curve above point b together with the recoil line a have been permanently wiped out as indicated by the dash lines in Figure 3.33.

A linear recoil line in a fixed excitation direction can be expressed in the scalar format as

$$B = \mu_r \mu_0 \left(H - H_c' \right),$$ (3.120)

where H_c', the coercivity of the linearized demagnetization curve, is termed recoil-line coercivity. For different recoil lines starting from different demagnetization points below the knee point K, the recoil-line coercivities are different.

For the vector format, even though the PM material might be isotropic, we still refer to the magnetization direction as the easy axis, and all directions orthogonal to the magnetization direction as the hard axes. In the hard axes, linear $B–H$ curve with the same permeability as that for the recoil lines are used. Assume $\mathbf{u}$ denotes the unit vector of the easy axis, then a recoil line can be expressed in the vector format as

$$\mathbf{B} = \mu_r\mu_0(\mathbf{H} - H_c'\mathbf{u}) \tag{3.121}$$

3.5.2 Irreversible Demagnetization Model

In the following illustration, the point (H, B) from the field solution is called the solution point, and the point located at the $B–H$ curve is termed working point. The iteration process is repeatedly using the parameters derived from the working point to solve the field equation, and determining the working point from the solution point until the total error between the solution point and the working point for all mesh elements satisfies the pre-specified condition.

In the FE implementation, two possible approaches can be applied to model permanent magnets with the nonlinear demagnetization behavior. The first approach is based on the magnetic field intensity (coercivity) H_c taken directly from the original demagnetization curve together with the treatment of anisotropic permeability. The coercivity H_c from the original demagnetization curve is treated as an additional source contributing to the right-hand side of the FE formulation. The original demagnetization curve in the easy axis is shifted to the first quadrant passing through the origin as a normal single-value nonlinear $B–H$ curve. In the other two directions orthogonal to the magnetization direction, a constant linear permeability with the value of the slope of the recoil line is applied. In such a case, the converged solution of Newton–Raphson iteration will guarantee the working point is on the original demagnetization curve (by shifting the working point back to the second quadrant). However, this approach is not suitable for modeling dynamic transient demagnetization behavior because the shifted demagnetization curve in the first quadrant as a normal $B–H$ curve has to be reconstructed if a new worst working point is found during the transient solution process (such as reconstructed from curve Br'-L_a-a-b-Hc to curve Br''-L_b-b-Hc in Figure 3.33). In addition, the Newton–Raphson iteration will not directly be applicable due to the discontinuity in the first derivative of the new constructed $B–H$ curve with a recoil line.

The second approach is based on a linearized model characterized by the recoil line passing through the latest identified demagnetization point, as illustrated below. Assume the point W_0 in Figure 3.34 is the worst working point so far during the process of transient FE analysis, and L_0 is the recoil line passing through point W_0. A new $B–H$ curve is constructed to consist of the recoil line L_0 and the partial demagnetization curve below point W_0, and is used for the new iteration.

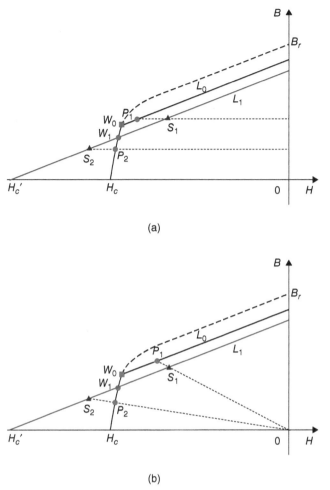

Figure 3.34   New working point determination in linearized demagnetization model: (a) B-correction scheme; (b) new scheme.

Also assume the working point of the last iteration is W_1, and its associated recoil line is L_1 with recoil-line coercivity of H_c', as shown in Figure 3.34. During the construction of field equations, the value of recoil-line coercivity H_c', instead of H_c directly from the original demagnetization curve, is used as an additional source contributing to the right-hand side of the FE formulation. At the same time, the linear permeability with the value of the slope of the recoil line is used for all (easy and hard) axes. If the solution point, obtained from the linear model based on H_c', is S_1 (or S_2), a new working point P_1 (or P_2) has to be identified on the constructed B–H curve for next field solution.

Since the recoil-line permeability, which is minimum differential permeability over whole demagnetization curve, is used during the nonlinear iteration, the

B-correction scheme of the fixed point method demonstrated in Section 3.4.3 can be employed for all PM mesh elements to guarantee convergence. However, the convergence rate could be very slow when the new working point is not located at the recoil line L_0. Hence, a new scheme is introduced to determine the new working point in each iteration to accelerate the convergence.

In the B-correction scheme, the new working point P_1 is identified by getting the H value from the constructed B–H curve based on the B value of the solution point S_1, or geometrically getting the intersection of a horizontal line passing through S_1, and intersecting with the constructed B–H curve, as shown in Figure 3.34a. In the new scheme, the new working point P_1 is obtained by getting the intersection of an auxiliary line passing through S_1 and the origin, and intersecting with the constructed B–H curve, as shown in Figure 3.34b.

The scalar field solution for $S_1(H_1, B_1)$ used in Figure 3.34 is

$$\begin{cases} H_1 = \mathbf{H}_1 \cdot \mathbf{u} \\ B_1 = \mathbf{B}_1 \cdot \mathbf{u} \end{cases} \tag{3.122}$$

The process of deriving new working point P_1 from the solution point S_1 will be repeated until the iteration is converged. After the nonlinear iteration has converged, this new working point W_1 (updated working point P_1 or P_2), if it is below point W_0, becomes the new worst working point for nonlinear iteration of next time step.

If the solution point S_1 is located in the third quadrant, the new working point P_1 is obtained directly based on the B-correction scheme, as shown in Figure 3.34a.

3.5.3 Temperature-Dependent Magnetic Properties

As most demagnetization curves are highly sensitive to temperature, a temperature increase can cause demagnetization [39]. To consider the temperature dependence of the demagnetization behavior, it is desirable to first describe the demagnetization curve by a function with a couple of temperature-dependent parameters as variables. These temperature-dependent parameters specified by supplier datasheets are normally associated with intrinsic flux density B_i vs H curve. Therefore, it is advantageous to work directly on intrinsic B_i–H curve, instead of B–H curve. Once the temperature-dependent B_i–H curve is derived, it is straightforward to convert B_i–H characteristic into B–H characteristic in terms of

$$B = B_i + \mu_0 H \tag{3.123}$$

If all B_i values of B_i–H curve are divided by B_r, and all H values are divided by H_{ci} which is termed intrinsic coercivity, this curve is called normalized B_i–H curve. A lot of supplier's demagnetization datasheets show that the normalized B_i–H curve almost does not depend on temperature. Therefore, a B_i–H curve at one temperature can be easily mapped to another temperature as long as the mapping functions from one temperature to another temperature are defined for B_r and H_{ci}. The mapping function for B_r and H_{ci} can be defined as

$$\begin{cases} B_r(T) = B_r(T_0)\,P(T) \\ H_{ci}(T) = H_{ci}(T_0)Q(T) \end{cases}, \tag{3.124}$$

where

$$\begin{cases} P(T) = 1 + \alpha_1(T - T_0) + \alpha_2(T - T_0)^2 \\ Q(T) = 1 + \beta_1(T - T_0) + \beta_2(T - T_0)^2 \end{cases} \tag{3.125}$$

with T_0 being the reference temperature, and α_1, α_2, β_1, and β_2 the coefficients provided in supplier datasheets.

Therefore, only the intrinsic demagnetization curve $B_i = f_{T0}(H)$ measured at temperature T_0, as well as parameters α_1, α_2, β_1, and β_2, is required from input. The intrinsic demagnetization curve at temperature T can be obtained as

$$\begin{aligned} B_i(H, T) &= f_{T0}(H/H_{ci}(T) \cdot H_{ci}(T_0))/B_r(T_0) \cdot B_r(T) \\ &= f_{T0}(H/Q(T)) \cdot P(T) \end{aligned} \tag{3.126}$$

After the B_i–H curve is derived from (3.126), the B–H curve in the second and third quadrants can be further derived through the conversion using (3.123).

The relative permeability of the recoil line at the reference temperature T_0 can be derived from (3.123) and (3.126) as

$$\begin{aligned} \mu_r(T_0) &= \frac{1}{\mu_0}\frac{\partial B(H, T_0)}{\partial H}\bigg|_{H=0} = \frac{1}{\mu_0}\frac{\partial B_i(H, T_0)}{\partial H}\bigg|_{H=0} + 1 \\ &= \frac{1}{\mu_0}f_{T0}{}'(0) + 1 \end{aligned} \tag{3.127}$$

Thus, the relative permeability of the recoil line at temperature T is

$$\begin{aligned} \mu_r(T) &= \frac{1}{\mu_0}\frac{\partial B_i(H, T)}{\partial H}\bigg|_{H=0} + 1 = \frac{P(T)}{Q(T)}\frac{1}{\mu_0}f_{T0}{}'(0) + 1 \\ &= \frac{P(T)}{Q(T)}\left(\mu_r(T_0) - 1\right) + 1 \end{aligned} \tag{3.128}$$

We observe from (3.128) that when $\mu_r(T_0) = 1$, $\mu_r(T) = 1$.

The validity of the presented temperature-dependent model can be illustrated by the example of the modeling of NdFeB N4517 magnet. Figure 3.35a is copied from the original vendor datasheet. The upper six curves are intrinsic B_i–H characteristics associated with different temperatures from 20°C to 120°C, and the lower six curves are corresponding normal B–H characteristics, respectively. On the other hand, as shown in Figure 3.35b, all curves can be adequately recovered based on the above-described temperature-dependent demagnetization model in terms of one set of discrete data describing single intrinsic B_i–H curve at temperature 20°C. It can be seen that each derived curve in Figure 3.35b matches extremely well with the corresponding curve on vendor datasheet in Figure 3.35a. In addition, in order to

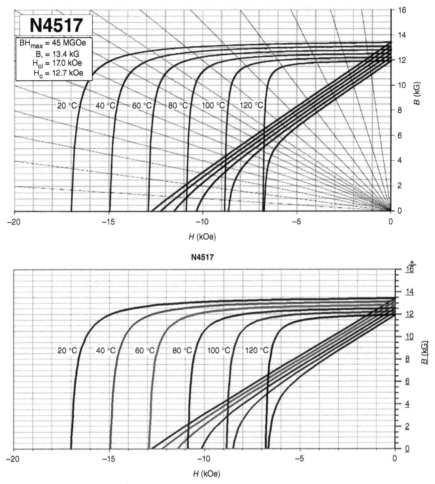

Figure 3.35   Comparison of *Bi-H* and *B–H* characteristics between (a) copied from vendor datasheet, and (b) derived based on the proposed temperature-dependent demagnetization model ($\alpha_1 = -0.1$, $\alpha_2 = 0$, $\beta_1 = -0.6$, and $\beta_2 = 0$).

account for possible severe demagnetization for practical applications, the derived *B–H* curves can be easily extended to the third quadrant, which is not shown in Figure 3.35b for the sake of comparison clarity.

   Figure 3.33 shows that when a new working point *a* goes below the knee point *K*, the subsequent working point will change along a new constructed *B–H* curve which consists of two curve segments: the recoil line L_a and the demagnetizing curve part below point *a*. The residual flux density of the recoil line is B_r', and the difference between B_r and B_r' is the irreversible B_r loss, which is denoted as $\Delta B_r = B_r - B_r'$. It is obvious that in the original demagnetizing curve, when the working point is above the knee point, the irreversible B_r loss is 0.

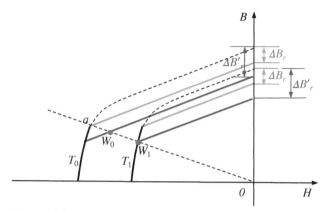

Figure 3.36   Irrevesible demagnetization when tempurature changes.

When the temperature of a PM with working point below the knee point changes from T_0 to T_1, the initial irreversible B_r loss ΔB_r keeps unchanged, based on which a new B–H curve can be constructed from the demagnetizing curve of the new temperature T_1. Assume the load line keeps unchanged, this new constructed B–H curve could be updated if a new worst working point W_1 causes the irreversible B_r loss to increase from ΔB_r to $\Delta B_r'$, see Figure 3.36. Now, if the temperature changes from T_1 back to T_0, the working point will not go back to point a, instead, it will locate at W_0, somewhere on the new constructed B–H curve based on the new irreversible B_r loss $\Delta B_r'$. A PM will remember the worst working point of all working temperatures.

3.5.4   Parameterized Demagnetization Curve

Many PM manufactures directly provide demagnetization curves for their products, but in most cases, manufactures provide only main parameters, such as residual flux density B_r, coercivity H_c, maximum energy product $(BH)_{max}$, and relative recoil permeability μ_r. In such a case, we need to construct the demagnetization curve based on these parameters.

Three-Parameter Algorithm
Given the three characteristic parameters B_r, H_c, and $(BH)_{max}$, the demagnetization curve can be geometrically constructed by the three-parameter algorithm, as shown in Figure 3.37.
    In Figure 3.37,

$$\begin{cases} H_a = H_c/a \\ B_a = B_r/a \end{cases},$$ (3.129)

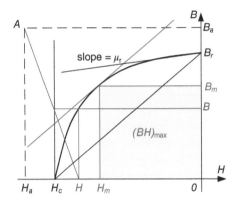

Figure 3.37   Three-parameter algorithm.

where

$$a = 2\sqrt{\left|\frac{B_r H_c}{(BH)_{\max}}\right|} - \left|\frac{B_r H_c}{(BH)_{\max}}\right| \tag{3.130}$$

For any magnetic field density H in the interval $[H_c, 0]$, flux density B can be expressed as

$$B = B_a \frac{H - H_c}{H - H_a} = B_r \frac{H - H_c}{aH - H_c} \tag{3.131}$$

The differential permeability is

$$\frac{dB}{dH} = B_r \frac{H_c(a - 1)}{(aH - H_c)^2} \tag{3.132}$$

The relative recoil magnetic permeability is

$$\mu_r = \frac{1}{\mu_0} \frac{dB}{dH}\bigg|_{H=0} = \frac{(a - 1)B_r}{\mu_0 H_c} \tag{3.133}$$

From the original demagnetization curve of Alnico 5, characteristic parameters are extracted and used to derive the constructed curve which is shown in Figure 3.38 compared with the original one for validation.

Four-Parameter Algorithm

Figure 3.38 shows that the three-parameter algorithm fits the demagnetization curve well. However, for the nonlinear permanent-magnetic material, the real working point lies normally not on the demagnetization curve, but on the recoil line. In three-parameter algorithm, the relative recoil permeability is obtained from other three parameters, see (3.133), which may have significant discrepancy with the

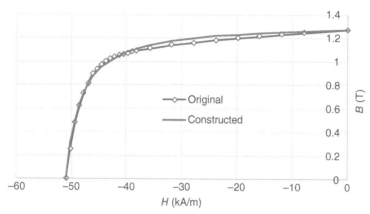

Figure 3.38   Constructed demagnetization curve for Alnico 5 based on three parameters $(B_r = 1.27$ T, $H_c = 50.9$ kA/m, $(BH)_{max} = 43.4$ kJ/m^3) compared with the original curve.

specified one. To force the relative recoil permeability match the specified value, four-parameter algorithm is introduced.

Based on four characteristic parameters B_r, H_c, $(BH)_{max}$, and μ_r, the four-parameter algorithm is summarized as:

1. Draw a line through the point $(0, B_r)$ with the slope equal to $\mu_r\mu_0$ as shown in Figure 3.39, the segment of this line in the second quadrant is termed ideal recoil line;

2. Find a virtual residual flux density B_{ro} so that the curve constructed by the three-parameter algorithm based on B_{r0}, H_c, and $(BH)_{max}$ will tangentially connect with the ideal recoil line.

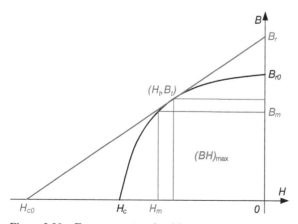

Figure 3.39   Four-parameter algorithm.

Flux density B as a function of H in interval $[H_c, 0]$ is

$$B = \begin{cases} B_{r0} \dfrac{H - H_c}{a_0 H - H_c} & H_c \leq H \leq H_t \\ B_r + \mu_r \mu_0 H & H_t \leq H \leq 0 \end{cases}, \tag{3.134}$$

where

$$a_0 = 2\sqrt{\left| \dfrac{B_{r0} H_c}{(BH)_{\max}} \right|} - \left| \dfrac{B_{r0} H_c}{(BH)_{\max}} \right| \tag{3.135}$$

Defining an initial interval $[B_0, B_1]$ with

$$\begin{cases} B_0 = \max(|\mu_r \mu_0 H_c|, |(BH)_{\max}/H_c|) \\ B_1 = B_r \end{cases} \tag{3.136}$$

the virtual residual flux density B_{r0} can be found by binary search as:

1. let $B_{r0} = (B_0 + B_1)/2$, get a_0 from (3.135);
2. let the differential permeability to be $\mu_r \mu_0$, get tangent-point field intensity H_t from (3.132), replacing a, B_r by a_0, B_{r0}, respectively;
3. calculate $dB_t = B_{t1} - B_{t2}$, where B_{t1} and B_{t2} are computed from (3.134) at H_t, respectively;
4. if $dB_t > 0$, the assumed B_{r0} is too large, let $B_1 = B_{r0}$; if $dB_t < 0$, the assumed B_{r0} is too small, let $B_0 = B_{r0}$;
5. repeat steps (1)~(4), until $|dB_t| < \varepsilon$, here ε is the specified error.

Figure 3.40 shows the constructed and original curves for Alnico 5.

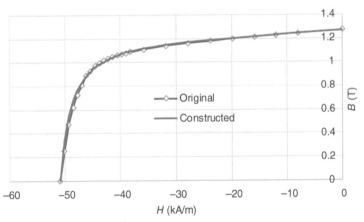

Figure 3.40 Constructed demagnetization curve for Alnico 5 based on four parameters ($B_r = 1.27$ T, $H_c = 50.9$ kA/m, $(BH)_{\max} = 43.4$ kJ/m^3, $\mu_r = 2.8$) compared with the original curve.

Four-Parameter Algorithm with Knee-Point Flux Density

For very hard PM materials with knee-point flux density $B_k < 0.5\ B_r$, the demagnetization curve cannot be constructed by the four-parameter algorithm presented above because the parameter $(BH)_{max}$ is not independent, which can be obtained from

$$(BH)_{max} = \frac{1}{4}\frac{B_r^2}{\mu_r\mu_0} \tag{3.137}$$

In such cases, a new parameter B_k, called knee-point flux density, instead of $(BH)_{max}$, can be introduced. The tangent connection point, see Figure 3.39, can be assumed to be at the knee point, that is

$$\begin{cases} B_t = B_k \\ H_t = (B_t - B_r)/(\mu_r\mu_0) \end{cases} \tag{3.138}$$

In (3.132), replace a, B_r by a_0, B_{r0}, respectively, and let the differential permeability at H_t be $\mu_r\mu_0$, then

$$B_{r0}\frac{H_c(a_0 - 1)}{(a_0H_t - H_c)^2} = \mu_r\mu_0 \tag{3.139}$$

Forcing flux density defined in (3.134) to continue at H_t, we get

$$B_{r0}\frac{H_t - H_c}{a_0H_t - H_c} = B_t \tag{3.140}$$

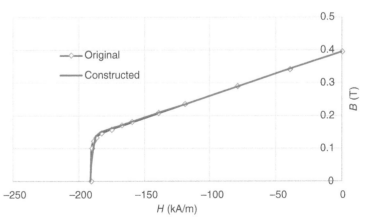

Figure 3.41   Constructed demagnetization curve for Ceramic 5 based on four parameters ($B_r = 0.395$ T, $H_c = 191$ kA/m, $B_k = 0.145$ T, $\mu_r = 1.068$) compared with the original curve.

Solving (3.140) together with (3.129), we obtain

$$
\begin{cases}
a_0 = \dfrac{B_r - \mu_r \mu_0 (H_t - H_c)}{B_r - \mu_r \mu_0 H_t (H_t / H_c - 1)} \\[4mm]
B_{r0} = B_t \dfrac{a_0 H_t - H_c}{H_t - H_c}
\end{cases}
\tag{3.141}
$$

With identified parameters in (3.141), we get flux density B for H in interval $[H_c, 0]$ by (3.134).

For the sake of validation, Figure 3.41 compares the constructed magnetization curve based on the knee-point flux density with the original curve for Ceramic 5.

REFERENCES

[1] T. N. T. Goodman, *Shape Preserving Representations, Mathematical Methods in Computer Aided Geometric Design (Oslo, 1988)*. Boston, MA: Academic Press, 1989, pp. 333–351. MR 1022717 (91a:65031).

[2] A. E. Umenei, Y. Melikhov, and D. C. Jiles, "Models for extrapolation of magnetization data on magnetic cores to high fields," *IEEE Trans. Magn.*, vol. 47, no. 12, pp. 4707–4711, 2011.

[3] G. Wolberg and I. Alfy, "An energy-minimization framework for monotonic cubic spline interpolation," *J. Comput. Appl. Math.*, vol. 143, pp. 145–188, 2002.

[4] M. H. Lam, "Monotone and convex quadratic spline interpolation," *Va. J. Sci.*, vol. 41, no. 1, pp. 3–13, 1990.

[5] A. D. Napoli and R. Paggi, "A model of anisotropic grain-oriented steel," *IEEE Trans. Magn.*, vol. 19, no. 4, pp. 1557–1561, Jul. 1983.

[6] J. M. Dedulle, G. Meunier, A. Foggia, and J. C. Sabonnadiere, "Magnetic fields in nonlinear anisotropic grain-oriented iron-sheet," *IEEE Trans. Magn.*, vol. 26, no. 2, pp. 524–526, Mar. 1990.

[7] J. Liu, A. Basak, A. J. Moses, and G. H. Shirkoohi, "A method of anisotropic steel modeling using finite element method with confirmation by experimental results," *IEEE Trans. Magn.*, vol. 30, no. 5, pp. 3391–3394, Sep. 1994.

[8] P. P. Silvester and R. P. Gupta, "Effective computational models for anisotropic soft B-H curves," *IEEE Trans. Magn.*, vol. 27, no. 5, pp. 3804–3807, Sep. 1991.

[9] D. Lin, P. Zhou, Z. Badics, W. N. Fu, Q. M. Chen, and Z. J. Cendes, "A new nonlinear anisotropic model for soft magnetic materials," *IEEE Trans. Magn.*, vol. 42, no. 4, pp. 963–966, Apr. 2006.

[10] G. Bertotti, "General properties of power losses in soft ferromagnetic materials," *IEEE Trans. Magn.*, vol. 24, no. 1, pp. 621–630, 1988.

[11] C. P. Steinmetz, "On the law of hysteresis," *AIEE Trans.*, vol. 9, pp. 3–64, 1892. Reprinted under the title "A Steinmetz contribution to the ac power revolution," Introduction by J. E. Brittain, in *Proc. IEEE*, vol. 72, pp. 196–221, Feb. 1984.

[12] D. Lin, P. Zhou, W. N. Fu, Z. Badics, and Z. J. Cendes, "A dynamic core loss model for soft ferromagnetic and power ferrite materials in transient finite element analysis," *IEEE Trans. Magn.*, vol. 40, no. 2, pp. 1318–1321, Mar. 2004.

[13] D. Lin, P. Zhou, Q. M. Chen, N. Lambert, and Z. J. Cendes, "The effects of steel lamination core losses on 3D transient magnetic fields," *IEEE Trans. Magn.*, vol. 46, no. 8, pp. 3539–3542, Aug. 2010.

[14] P. Zhou, D. Lin, C. Lu, N. Chen, and M. Rosu, "A new algorithm to consider the effects of core losses on 3-D transient magnetic fields," *IEEE Trans. Magn.*, vol. 50, no. 2, Article 7008904, Feb. 2014.

[15] P. Zhou, W. N. Fu, D. Lin, S. Stanton, and Z. J. Cendes, "Numerical modeling of magnetic devices," *IEEE Trans. Magn.*, vol. 40, no. 4, pp. 1803–1809, Jul. 2004.

[16] Z. Ren, "T-Ω formulation for eddy-current problems in multiply connected regions," *IEEE Transactions on Magnetics*, vol. 38, no. 2, pp. 557–560, Mar. 2002.

[17] I. D. Mayergoyz, *Mathematical Models of Hysteresis*. New York: Springer Verlag, 1991.

[18] E. Cardelli, E. Della Torre, and A. Faba, "A general vector hysteresis operator: Extension to the 3-D case," *IEEE Trans. Magn.*, vol. 46, no. 12, pp. 3990–4000, 2010.

[19] E. Cardelli, "A general vector hysteresis operator for the modeling of vector fields," *IEEE Trans. Magn.*, vol. 47, no. 8, pp. 2056–2067, 2011.

[20] A. Bergqvist, "Magnetic vector hysteresis model with dry friction-like pinning," *Physica B*, vol. 233, no. 4, pp. 342–347, 1997.

[21] M. d'Aquino, C. Serpico, C. Visone, and A. A. Adly, "A new vector model of magnetic hysteresis based on a novel class of play hysterons," *IEEE Trans. Magn.*, vol. 39, no. 5, pp. 2537–2539, 2003.

[22] J. V. Leite1, N. Sadowski1, P. A. D. Silva, Jr., N. J. Batistela1, P. Kuo-Peng1, and J. P. A. Bastos, "Modeling magnetic vector hysteresis with play hysterons," *IEEE Trans. Magn.*, vol. 43, no. 4, pp. 1401–1404, 2007.

[23] T. Matsuo and M. Miyamoto, "Dynamic and anisotropic vector hysteresis model based on isotropic vector play model for nonoriented silicon steel sheet," *IEEE Trans. Magn.*, vol. 48, no. 2, pp. 215–218, 2012.

[24] E. Della Torre, *Magnetic Hysteresis*. Piscataway, NJ: IEEE Press, 2000.

[25] C. Visone, D. Davino, and A. A. Adly, "Vector Preisach modeling of magnetic shape memory materials oriented to power harvesting applications," *IEEE Trans. Magn.*, vol. 46, no. 6, pp. 1848–1851, 2010.

[26] G. R. Kahler, E. Della Torre, and E. Cardelli, "Implementation of the Preisach-Stoner-Wohlfarth classical vector model," *IEEE Trans. Magn.*, vol. 46, no. 1, pp. 21–28, 2010.

[27] D. Lin, P. Zhou, and A. Bergqvist, "Improved vector play model and parameter identification for magnetic hysteresis materials," *IEEE Trans. Magn.*, vol. 50, no. 2, Article 7008704, 2014.

[28] M. Brokate, "Some mathematical properties of the Preisach model for hysteresis," *IEEE Trans. Magn.*, vol. 25, no. 4, pp. 2922–2924, Jul. 1989.

[29] E. Dlala, A. Belahcen, K. A. Fonteyn, and M. Belkasim, "Improving loss properties of the Mayergoyz vector hysteresis model," *IEEE Trans. Magn.*, vol. 46, no. 3, pp. 918–924, 2010.

[30] F. Hantila, G. Preda, and M. Vasiliu, "Polarization method for static fields," *IEEE Trans. Magn.*, vol. 36, no. 4, pp. 672–675, Jul. 2000.

[31] E. Dlala, A. Belahcen, and A. Arkkio, "Locally convergent fixed-point method for solving time-stepping nonlinear field problems," *IEEE Trans. Magn.*, vol. 43, no. 11, pp. 3969–3975, Nov. 2007.

[32] E. Dlala and A. Arkkio, "Analysis of the convergence of the fixed-point method used for solving nonlinear rotational magnetic field problems," *IEEE Trans. Magn.*, vol. 44, no. 4, pp. 473–478, Apr. 2008.

[33] M. E. Mathekga, R. A. McMahon, and A. M. Knight, "Application of the fixed point method for solution in time stepping finite element analysis using the inverse vector Jiles-Atherton model," *IEEE Trans. Magn.*, vol. 47, no. 10, pp. 3048–3051, Oct. 2011.

[34] W. Li and C. Koh, "Investigation of the vector Jiles-Atherton model and the fixed point method combined technique for time periodic magnetic problems," *IEEE Trans. Magn.*, DOI: 10.1109/TMAG.2014.2360150.

[35] P. Zhou, D. Lin, B. He, S. Kher, and Z. J. Cendes, "Strategy for accelerating nonlinear convergence for T-Ω formulation," *IEEE Trans. Magn.*, vol. 46, no. 8, pp. 3129–3132, Aug. 2010.

[36] S. Ruoho, J. Kolehmainen, J. Ikaheimo, and A. Arkkio, "Interdependence of demagnetization, loading and temperature rise in a permanent-magnet synchronous motor," *IEEE Trans. Magn.*, vol. 46, no. 3, pp. 949–953, Mar. 2010.

[37] S. Ruoho, E. Dlala, and A. Arkkio, "Comparison of demagnetization models for finite-element analysis of permanent-magnet synchronous machines," *IEEE Trans. Magn.*, vol. 43, no. 11, pp. 3964–3968, Nov. 2007.

[38] P. Zhou, D. Lin, Y. Xiao, N. Lambert, and M. A. Rahman, "Temperature-dependent demagnetization model of permanent magnets for finite element analysis," *IEEE Trans. Magn.*, vol. 48, no. 2, pp. 1031–1034, Feb. 2012.

[39] A. Bergqvist, D. Lin, and P. Zhou, "Temperature-dependent vector hysteresis model for permanent magnets," *IEEE Trans. Magn.*, vol. 50, no. 2, Article 7008404, Feb. 2014.

CHAPTER 4

THERMAL PROBLEMS IN ELECTRICAL MACHINES

4.1 INTRODUCTION

Thermal analysis of electric motors is in general regarded as a more challenging area of analysis than electromagnetic analysis in terms of the difficulties in constructing a model and achieving good accuracy. This is due to several factors:

i. Many electric motor electromagnetic problems can be reduced to a two-dimensional (2D) problem which can be fully described using a simple set of analytical equations, or Maxwell's equations when using numerical finite element analysis. The thermal analysis of an electric motor is seen to be more of a three-dimensional (3D) problem, with complex heat transfer phenomena to solve such as heat transfer through complex composite components like the wound slot, temperature drop across interfaces between components, and complex turbulent air flow within the endcaps around the end winding that includes rotational effects.

ii. Most motor designers are from an electrical background and in most electrical engineering degree courses, thermal analysis, which is a mechanical engineering discipline, is only given a rudimentary treatment. Also, emphasis is often given to complex thermodynamics theory and topics relevant to power stations such as steam tables rather than simple heat transfer analysis based on conduction, convection, and radiation thermal resistances.

Experts in heat transfer analysis can often baffle an electromagnetic engineer in jargon associated with thermal analysis, that is, ask about the magnitude of Reynolds number (Re), to try and judge if there is laminar or turbulent flow, and look at interface resistances in units of K/W/m^2. Conversely, the electromagnetic expert can baffle the heat transfer expert with talks about limiting flux densities, MMF harmonics, etc.

Heat transfer is not too complex if we simplify the mathematical concepts and formulate a heat transfer language understandable by non-specialists using parameters that are easy to visualize like effective gaps in mm rather than interface resistance

Multiphysics Simulation by Design for Electrical Machines, Power Electronics, and Drives, First Edition.
Marius Rosu, Ping Zhou, Dingsheng Lin, Dan Ionel, Mircea Popescu, Frede Blaabjerg, Vandana Rallabandi, and David Staton.

in $K/W/m^2$. Also, to show physical quantities for a particular design and fluid type, like graphs of surface dissipation (heat transfer coefficient in $W/m^2/K$) against fluid velocity, rather than the equivalent dimensionless graph of Nusselt number (Nu) against Re number. As electrical engineers have a good knowledge of electrical resistance networks, the thermal resistance network is usually easy to understand. The temperature is the equivalent of the voltage, the power the equivalent of the current, and the thermal resistance the equivalent of the electrical resistance. There are only three types of thermal resistance for conduction, convection, and radiation. In modern simulation tools [4] the thermal resistance network is automatically constructed from the definitions of the geometry, materials, and cooling type. A user interface can be provided to make data input easy, that is, using parameterized cross-section editors, material libraries and many pre-setup editors to help in setting data for the different cooling options with inputs of meaningful quantities such as water jacket flow rates.

Losses are induced in different components of the machine such as stator iron, windings, rotor cages and magnets, and there needs to be sufficient dissipation to minimize thermal stress. We can thermally protect electrical machines by reducing local losses, that is, the induced eddy current losses in the electrical conducting regions, iron cores, magnets, retaining sleeves; and/or using an efficient cooling system. Depending on the application, cooling systems can be employed with natural convection (totally enclosed non-ventilated, TENV), forced convection (air or liquid cooling), or, in the case of electrical machines, operating in a vacuum, with radiation cooling.

The thermal analysis of an electric motor is generally regarded as more challenging in comparison with the electromagnetic analysis, in terms of the ease of construction of a model and achieving good accuracy. This is because it is highly dependent not only on the design but on the manufacturing tolerances.

Before the advent of computers, motor sizing was traditionally made using so-called D^2L, D^3L and D^xL equations, where D stands for the air-gap diameter and L is the axial active length of the machine. The designer is providing limiting values of specific magnetic loading and specific electric loading and/or current density from previous experience. This method of sizing does not involve thermal analysis directly, the specific electric loading and current density being limited to prevent overheating. At that time, a simple thermal network analysis based on lumped parameters was also used by some designers to perform rudimentary thermal analysis, but the networks were kept as simple as possible so they could be calculated by hand, for example, maybe just one thermal resistance to calculate the steady-state temperature rise of the winding. With the introduction of computers to electrical motor design process, the complexity of the thermal networks has increased. A reference paper highlighting the introduction of more complex thermal networks calculated using computers was published in 1991 [16]. Thermal network analysis has become the main tool used by many researchers involved in thermal analysis of electrical machines, both for steady-state and transient analyses. Another factor that has led to increased interest in thermal network analysis was the introduction of induction motor inverter supplies. Several authors have studied the effect of increased losses, resulting from Six-Step and PWM voltages, on motor temperatures.

Thermal analysis has always received less attention than electromagnetic design in electrical machines analysis. However, in twenty-first century the topic had started to receive larger importance due to market globalization, and the requirement for smaller, cheaper, and more efficient electric motors. In many cases, software used for design of electric machines has now adopted improved thermal modeling capabilities and features enabling better integration between the electromagnetic and thermal design.

Several interesting studies have been published in recent years on thermal analysis of electric machines: coupled electromagnetic and thermal analysis with the thermal network solved using network analysis; the losses are calculated using analytical methods or electromagnetic finite element analysis is used. A thermal network method is proposed to account for combined air flow and heat transfer, that is, for forced air cooling in stator and rotor core ducts in this case. We can combine network and computational fluid dynamics (CFD) method to model thermal aspects in electrical machine. Network analysis is used to calculate conduction through the electromagnetic structure while CFD is used for convection at the surface. The use of CFD for prediction of convective heat transfer is expanded further in this chapter. Calibration with measured data is typically used to calibrate thermal resistances that are influenced by the manufacturing process of the motor. An example is the thermal interface resistance between stator lamination and housing which is influenced by the method used to insert the stator in the frame.

4.2 HEAT EXTRACTION THROUGH CONDUCTION

Conduction heat transfer mode is created by the molecule vibration in a certain material [1–4]. Typically, a material with good electrical conductivity is also characterized by good thermal conductivity. In an electrical machine, it is also desirable to have materials that are good electrical insulators and have good thermal conductivity.

The thermal conduction phenomenon will depend on the thermal conductivity (k) and dimensions (length L and area A) of the region, with the thermal resistance given by:

$$R_{th} = \frac{L}{kA} \tag{4.1}$$

Equation (4.1) shows that a good heat extraction through conduction, that is, low thermal resistance of a motor component, is achieved either if the material has a high thermal conductivity or if the ratio of the length over the area is minimized. The latter is obtained, for example, by having short active axial length or end windings and high slot fill factor which means packing as much wire material as technically possible into the slots of the electrical machine.

Metals, that is, copper, aluminum, steel, have high thermal conductivity due to their well-ordered crystalline structure. The thermal conductivity for metals is usually in the range 15–400 W/m/K. Solid insulator materials, that is, resins, paper, do not have a well-ordered crystalline structure and are often porous. Thus, the energy transfer between molecules is impeded. The thermal conductivity for insulators is typically in the range 0.1–1 W/m/K. For comparison, air has an average thermal conductivity of 0.026 W/m/K.

The last two decades have seen significant research efforts taking place in the development of insulation materials that have a higher thermal conductivity value. Table 4.1 summarizes the actual standard insulation classes as approved by IEC and NEMA. Notice that an existing insulation class at 300°C is not part of the standard. In Table 4.2, some of the latest high thermal conductivity insulation materials with high volume electric resistivity are reported. Figures 4.5 and 4.6 show various solutions to achieve high winding slot fill factors. Most of these are suitable for electrical machines with concentrated fractional slots/pole configurations, for example, Slots = Poles ± 2; Poles ± 3; etc. Such topologies can be built with pre-formed tooth wound coils. When the coils are pre-formed, various shapes can be considered and the conductors can be subjected to a high pressure assembly method that would increase significantly the amount of conducting material, copper or aluminum, that can be packed into the machine's slots. Other AC electrical machine types, such as induction or synchronous wound field, cannot benefit from these solutions as they generally require a distributed winding configuration.

In an alternative current (AC) field, any electrical machine may benefit from using aluminum magnet wire. This is valid in special applications like automotive and aerospace, where a high rotational speed (>10,000 rpm) and highly transient duty cycles might be required. For comparison, the two main materials used for windings, that is, copper and aluminum properties, are given in Table 4.3.

In high speed applications, the winding losses will comprise a DC component and an AC component. If the induced eddy currents are resistance limited and the magnetic field generated by the eddy currents is negligible compared to the external field, simplified formulae can be deduced to estimate the specific AC losses in the winding (equations 4.2 and 4.3). Notice that the inductance of the eddy current path and the mutual interactions between neighboring conductors are not catered for. This approach is accurate where the conductor dimension is small relative to the skin depth effect of the magnetic field variation and has been successfully applied to evaluating eddy current losses in randomly disposed, multi-stranded electrical machine windings.

Round conductor specific AC losses:

$$P_{ac} = \frac{\pi l d_c^4 (\omega B)^2}{64\,\rho} \qquad (4.2)$$

Rectangular conductors specific AC losses:

$$P_{ac} = \frac{l h_c w_c^3 (\omega B)^2}{12\,\rho} \qquad (4.3)$$

where ρ is the electrical resistivity, l is the conductor active length, B is the average value of a sinusoidal external field, and ω is the electric pulsation. The rectangular conductor cross-section is defined by its height h_c and width w_c, and the circular conductor by its diameter d_c.

TABLE 4.1 Standard insulation classes [5]

IEC 60085 Thermal class [2]	Old IEC 60085 Thermal class [2]	NEMA class [3]	NEMA/UL letter class [3]	Maximum hot spot temperature allowed	Typical materials
70					
90	Y			90°C	Unimpregnated paper, silk, cotton, vulcanized natural rubber, thermoplastics that soften above 90°C [4]
105	A	105	A	105°C	Organic materials such as cotton, silk, paper, some synthetic fibers [5]
120	E			120°C	Polyurethane, epoxy resins, polyethylene terephthalate, and other materials that have shown a usable lifetime at this temperature
130	B	130	B	130°C	non-organic materials such as mica, glass fibers, asbestos, with high-temperature binders or others with usable lifetime at this temperature
155	F	155	F	155°C	Class 130 materials with binders stable at the higher temperature or other materials with usable lifetime at this temperature
180	H	180	H	180°C	Silicone elastomers and class 130 non-organic materials with high-temperature binders, or other materials with usable lifetime at this temperature
200			N	200°C	As for class B and including Teflon
220		220	R	220°C	As for IEC class 200
		240	S	240°C	Polyimide enamel (Pyre-ML) or polyimide films (Kapton and Alconex GOLD)
250				250°C	As for IEC class 200. Further IEC classes designated numerically at 25°C increments
N/A	N/A	N/A	N/A	300	Newly developed polymer (NeoTem^{TM}) at Zeus Inc.

TABLE 4.2 Insulation materials

Material	Thermal conductivity (W/m/K)	Electric resistivity (ohm × cm)	Manufacturer
MC4260	0.60–0.70	8×10^{14}	Elantas
Kapton FN	0.12	$1.4–2.3 \times 10^{17}$	DuPont
Nomex paper	0.12–0.15	$8 \times 10^{11}–8 \times 10^{16}$	DuPont
PEEK 450G	0.25	4.9×10^{16}	Victrex
Mylar*	0.140	$10^{13}–10^{18}$	DuPont
ECCTreme ECEA 3000	0.180	$>10^{18}$	DuPont
Teflon PTFE	0.22	$>10^{18}$	DuPont

The AC losses will decrease if the electrical resistivity of the material is higher and therefore the copper made windings will exhibit higher AC losses than aluminum made winding for the same dimensions, and load. An increased ratio of conducting material area over the slot area can be achieved by using: pre-formed bobbins, Figure 4.1; profiled, flat wire, Figure 4.2.

4.3 HEAT EXTRACTION THROUGH CONVECTION

Convection heat transfer mode appears between a surface and a fluid due to intermingling of the fluid immediately adjacent to the surface where a conduction transfer mode will occur with the remainder of the fluid due to the molecules' motion [4, 7–11].

Generically, we distinguish between:

a. natural convection when the fluid motion is due to buoyancy forces that arise from the change in density of the fluid in the vicinity of a surface; and

b. forced convection when the fluid motion is due to an external force created by a special device, for example, fans, pumps.

Based on the fluid flow type, it is possible to have laminar flow (this is a streamlined flow and occurs at lower velocity) and a turbulent flow. Turbulent flow is created by eddies at higher velocities resulting in an enhanced heat transfer by comparison with the laminar flow case, but with a larger pressure drop. Convection depends on

TABLE 4.3 Copper and aluminum wire properties

Material	Electric resistivity at 20°C (ohm × m)	Thermal conductivity (W/m/K)	Density (kg/m^3)	Specific heat (kJ/kg/K)
Copper	1.724×10^{-8}	386	8890	0.385
Aluminum 99.9%	2.826×10^{-8}	205	2700	0.833

Figure 4.1   Pressed coil for high slot fill factor.

the heat transfer coefficient h that is determined empirically from test data or from CFD analysis.

The convection heat transfer can be computed as:

$$Q = h_c A (T_{\text{wall}} - T_{\text{fluid}}) \tag{4.4}$$

The convection thermal resistance is calculated using:

$$R_{\text{th}} = \frac{1}{hA} \tag{4.5}$$

where

h_c = convection heat transfer coefficient (W/m^2/K)

A = wall surface area (m^2)

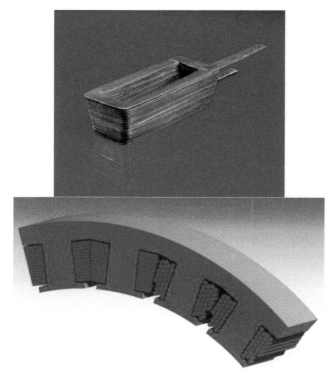

Figure 4.2   Profiled and standard coils [6]. *Source*: http://www.ifam.fraunhofer.de/

Convection depends on the heat transfer coefficient h_c that is determined empirically from test data or from CFD analysis.

A set of dimensionless numbers can be used to define in detail the convection phenomenon. The dimensionless numbers are functions of fluid properties, size (characteristic length), fluid velocity (forced convection), temperature (natural convection), and gravity (natural convection).

- Reynolds number (R_e)—inertia force/viscous force

$$R_e = \rho v L / \mu \tag{4.6}$$

- Grashof number (G_r)—buoyancy force/viscous force

$$G_r = \beta g \theta \rho^2 L^3 / \mu^2 \tag{4.7a}$$

$$Gr = \frac{g \beta \rho^2 (T_{\text{wall}} - T_{\text{fluid}}) L^3}{\mu^2} \tag{4.7b}$$

- Prandtl number (P_r)—momentum/thermal diffusivity for a fluid

$$P_r = c_p \mu / k \tag{4.8}$$

- Nusselt number (N_u)—convection/conduction heat transfer in a fluid

$$N_u = hL/k \qquad (4.9)$$

where

h = heat transfer coefficient (W/m^{2}C)
μ = fluid dynamic viscosity (kg/s m)
ρ = fluid density (kg/m^3)
k = thermal conductivity of the fluid (W/mC)
c_p = specific heat capacity of the fluid (kJ/kg C)
v = fluid velocity (m/s)
θ = surface to fluid temperature (°C)
L = characteristic length of the surface (m)
β = coefficient of cubical expansion of fluid (1/C)
g = acceleration due to gravity (m/s^2)

As convective heat transfer is non-dimensionalized with Nusselt number (Nu), h_C can be calculated using empirical correlations based on dimensionless numbers (Re, Gr, Pr):

$$Nu = \frac{h_c L}{k} = f\,(Re,\ Gr,\ Pr) \qquad (4.10)$$

- L = characteristic length (m)
- k = fluid thermal conductivity (W/m/K)

The dimensionless numbers allow the same formulation to be used with different fluids, dimensions, and models of dynamic and geometric similarity to those used in the original experiments. A lot of correlations are available in the literature and specialized software as Motor-CAD automatically selects the most appropriate correlation that matches the cooling type and geometry shape: cylinder, flat plate, open/enclosed channel, etc.

4.3.1 Natural Convection

There is a general form of natural convection correlation:

$$Nu = f\,(Gr, Pr) = C(Gr\ Pr)^n \qquad (4.11)$$

- C and n are curve fitting constants
- Rayleigh number, $Ra = Gr\ Pr$
- Transition from laminar to turbulent flow: $10^7 < Ra < 10^9$
- All fluid properties are evaluated at the film temperature,

$$T_f = \frac{(T_{\text{wall}} + T_{\text{fluid}})}{2} \qquad (4.12)$$

4.3.2 Forced Convection

General form of forced convection correlation:

$$Nu = f(Re, Pr) = CRe^m Pr^n \qquad (4.13)$$

C, m, and n are curve fitting constants.
Re is a measure of inertial forces to viscous forces.
For external flow, laminar/turbulent transition $Re \approx 5 \times 10^5$
For internal flow, the following observations are valid:

- The flow is assumed to be fully laminar when $Re < 2300$ in circular/rectangular channels and when $Re < 2800$ in concentric cylinders.
- The flow is assumed to be fully turbulent when $Re > 5000$ (in practice the flow may not be fully turbulent until $Re > 10,000$).
- A transition between laminar and turbulent flow is assumed for Re values between those given above.

4.3.3 Enclosed Channel Forced Convection

Laminar Flow
Concentric cylinders (adaptation of formulation for parallel plates which includes entrance length effects):

$$Nu = 7.54 + \frac{\left[0.03 \times {D_h}/{L} \times Re\,Pr\right]}{\left[1 + 0.016 \times \left({D_h}/{L} \times Re\,Pr\right)^{\frac{2}{3}}\right]} \qquad (4.14)$$

The second term in the above equation is the entrance length correction which accounts for entrance lengths where the velocity and temperature profiles are not fully developed.

Circular channels (which includes entrance length effects):

$$Nu = 3.66 + \frac{\left[0.065 \times {D_h}/{L} \times Re\,Pr\right]}{\left[1 + 0.04 \times \left({D_h}/{L} \times Re\,Pr\right)^{\frac{2}{3}}\right]} \qquad (4.15)$$

Rectangular channels (adaptation of formulation for round channels):

$$Nu = 7.46 - 17.02\frac{H}{W} + +22.43\left(\frac{H}{W}\right)^2 - 9.94\left(\frac{H}{W}\right)^3$$
$$+ \frac{\left[0.065 \times {D_h}/{L} \times Re\,Pr\right]}{\left[1 + 0.04 \times \left({D_h}/{L} \times Re\,Pr\right)^{\frac{2}{3}}\right]} \qquad (4.16)$$

where H/W = channel height/width ratio.

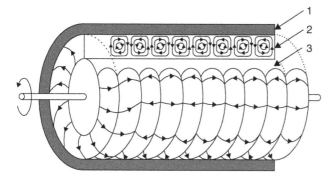

1: Stationary outer cylinder

2: Taylor vortices

3: Rotating inner cylinder

Figure 4.3   Turbulent flow in [8]. Reproduced with permission of Royal Society.

Turbulent Flow

We can use Gnielinski's formula for fully developed turbulent flow ($3000 < Re < 1 \times 10^6$)

$$Nu = \frac{(f/8)\,(Re - 1000)\,Pr}{1 + 12.7(f/8)^{1/2}(Pr^{2/3} - 1)}, \tag{4.17}$$

where f is Darcy friction factor and for a smooth wall is:

$$f = [0.79\ln{(Re)} - 1.64]^{-2} \tag{4.18}$$

In cooling channels, the fluid boundary layer is developing at the duct entrance before it becomes fully developed. A correction factor is applied for the developing flow.

It is noted that h increases dramatically as the flow regime changes from being laminar to turbulent flow, Figures 4.3 and 4.4.

For transitional flow heat transfer, Nu is calculated for both laminar and turbulent flow using the above formulations. A weighted average (based on Re) is then used to calculate Nu in the transition zone.

Modern applications where electrical machines are employed frequently rely on forced convection cooling systems that use air or liquid, that is, water, oil, and their combinations. Based on the convection technique, we have: air natural convection ($h = 5$–10 W/m^2/K), air forced convection ($h = 10$–300 W/m^2/K), and liquid forced convection ($h = 50$–20000 W/m^2/K).

Forced convection heat transfer from a given surface is a function of the local flow velocity. In order to predict the local velocity, a flow network analysis is performed to calculate the fluid flow (air or liquid) through the machine. Empirical dimensionless analysis formulations are used to predict pressure drops for flow restrictions such as vents, bends, contractions, and expansions. The governing equation that relates pressure drop (P (Pa), equivalent to an electrical voltage) to volume

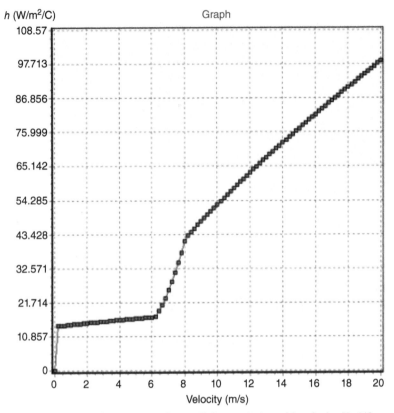

Figure 4.4   Convection heat transfer coefficient variation with velocity [9–11]

flow rate (Q (m³/s), equivalent to electrical current) and fluid dynamic resistance (R (kg/m⁷)) is:

$$P = R \cdot Q^2 \qquad (4.19)$$

The formulation is in terms of Q^2 rather than Q due to the turbulent nature of the flow. Two types of flow resistance exist. First, where there is a change in flow condition, such as expansions, contractions, and bends. Second, due to fluid friction at the duct wall surface: in electrical machines, this is usually negligible compared to the first resistance type due to the comparatively short flow paths. The flow resistance is calculated for all changes in the flow path using (4.20).

$$R = \frac{K \cdot d}{2 \cdot A^2} \qquad (4.20)$$

where d (kg/m³) is the air density, A (m²) is the flow area, and K is the dimensionless coefficient of local fluid resistance whose value depends upon the local flow condition (obstruction, expansion, contraction, etc.). Many empirical formulations are available in the technical literature to calculate the K factor for all changes in flow section within the motor. A merit of thermal tool is to automatically select the most appropriate formulation for all the flow paths involved, that is, a sudden contraction when air enters the stator/rotor ducts, a 90° bend where the air passes around the end winding, etc.

Forced convection can be achieved using configurations of channels, ducts, water jackets, spray cooling, and axle cooling:

i. TENV motors with various housing design types; for this cooling type, the heat extraction strongly depends on the motor dimensions, the viscous force, and the thermal diffusivity of the fluid.

ii. Fan cooled motors (TEFC) in which the fin channel design is essential, typically, an external fan is used to blow a fluid (normally air) across the outside of the machine. The heat dissipation from the outside of the motor is assumed to be a combination of mixed convection (forced convection due to fan combined with natural convection), radiation, and conduction. The cooling fluid can be air or any other fluid.

iii. Though ventilation with rotor and stator cooling ducts is characterized by a flow of fluid (usually air) through three parallel paths (stator axial ducts, rotor axial ducts, and airgap) of the machine, the fluid passing through the machine can be air or any other fluid like water or oil.

iv. Water jackets with various design types (axial and circumferential ducts) and stator and rotor water jackets.

v. Submersible cooling in which the motor is operating in a liquid environment such as water, oil, or a mixture of water, oil, and air.

vi. Wet rotor and wet stator cooling is where a fluid is passed down the airgap of the machine and over the shaft and end-winding surfaces in the endcap regions of the machine.

vii. Spray cooling, when the cooling fluid is passed down a hole on the shaft and is sprayed directly at the inner surface of the end windings due to the centrifugal forces.

viii. Direct conductor cooling, for example, slot water jacket when a cooling fluid is flowing through the slots; it is possible to have internal ducts within the slot to contain the fluid, or the fluid is just completely flowing between conductors.

Table 4.4 gives the thermal and mechanical properties for the most used fluids in forced convection systems for power traction motors.

Figure 4.5 shows a TENV servomotor that has improved heat extraction by optimizing the space and shape of the housing fins.

In Figure 4.6, an in-wheel motor with natural convection air cooled rotor and axle water jacket with EWG50/50 cooling fluid is presented. The central axle of the wheel acts also as a large heat sink.

Figure 4.7a illustrates the Nissan Leaf electric motor assembly. This is a brushless permanent magnet (BPM) motor that is cooled via a water jacket with three parallel paths and using as cooling fluid EGW 50/50, with 6 l/min flow rate and 65°C

TABLE 4.4   Cooling fluids properties—average values at 0°C–40°C

Fluid	Thermal conductivity (W/m/K)	Specific heat (kJ/kg/K)	Density (kg/m^3)	Kinematic viscosity (m^2/s)
Air (sea level)	0.0264	1.0057	1.1174	1.57×10^{-5}
Brayco Micronic	0.1344	1.897	835	1.35×10^{-5}
Dynalene HF-LO	0.1126	2.019	778	3.2×10^{-6}
EGW 50/50	0.37	3.0	1088	7.81×10^{-6}
EGW 60/40	0.34	3.2	1100	1.36×10^{-5}
Engine oil	0.147	1.796	899	4.28×10^{-3}
Mobil Jet Oil	0.149	1.926	1014	1.88×10^{-4}
Paratherm LR	0.1532	1.925	778	3.43×10^{-6}
PGW 50/50	0.35	3.5	1050	1.9×10^{-5}
PGW 60/40	0.28	3.25	1057	3.31×10^{-5}
RF 245 FA	0.014	0.9749	10.51	1.027×10^{-5}
Silicone KF96	0.15	1.5	1000	8×10^{-5}
Skydrol 500-4	0.1317	1.75	1000	3.5×10^{-5}
Water	0.56	4.217	1000	1.78×10^{-6}

inlet temperature. In Figure 4.7b, the modeled temperature distribution in the water jacket is given.

Figures 4.8 and 4.9 present a new high torque density motor for electric racing cars. With all the electromagnetic loss components minimized, an extremely efficient cooling system is still required for the heat extraction from the motor without compromising the overall system performance. The stator winding is cooled with forced liquid—Paratherm LR, Table 4.4—through the slot.

The fluid is retained in the stator slots region with a carbon fiber tube, whilst the end windings are potted. The volume flow rate is variable between 5 and 15 l/min allowing a winding temperature below 120°C in all cases.

In the airgap, a constant air volume flow rate of 1000 l/min helps with maintaining the rotor surface and magnet's temperature at less than 160°C. Inlet temperatures are between 40°C and 50°C.

Figure 4.10 shows YASA cooling system. The stator is formed by individual teeth carrying individual coils and assembled together using resins, so that there is no back iron—yokeless armature topology.

There is a water jacket type housing that contains cooling liquid. The stator housing has heat dissipating fins accessible to the open environment whereby air movement relative to the housing, caused by rotation of the rotor, absorbs heat from the fins. The motor is an axial flux machine and is used for in-wheel vehicle applications.

Another recently popular and efficient cooling system is to implement oil spray cooling, with or without the use of nozzles. A typical arrangement is to fill the motor partially with oil such that the oil in contact with the rotor splashes around the endspace and cools the surfaces that it splashes over. Such a method is used in the Toyota

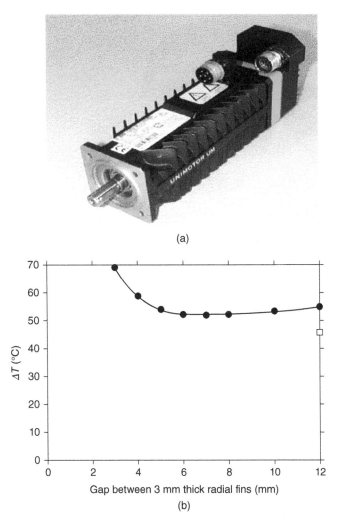

(a)

(b)

Figure 4.5   (a) Servomotor with optimized fins for natural convection, (b) influence of the gap between fins on motor thermal response [19, 21].

Prius traction electric motor. The level of oil fill is important in this case. Too little oil and there will be no pick up of oil from the rotor and so no splashing. Too much oil will hinder the splashing and also increase the windage loss due to oil in the airgap.

Yet another arrangement is to feed oil onto the rotor such that it is thrown off the axial ends by rotation so hitting the end windings and other surfaces in the end-space. A good cooling fluid for electrical machines using the spray method should have acceptable properties as listed: chemically stable and inert, non-toxic, non-flammable, low dielectric constant, high dielectric strength, high electric resistivity. Obviously, water is not suitable due to its properties: electrically conductive, corrosive, etc.

Figure 4.6 In-wheel electric motor, air cooled and axle water jacket [66]. Reproduced with permission of Protean Electric Ltd.

(a)

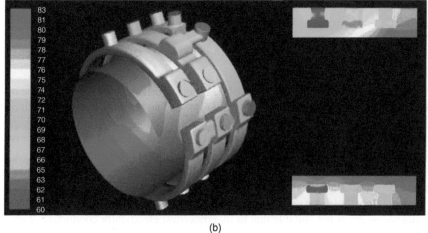

(b)

Figure 4.7 (a) Nissan leaf electric motor; (b) CFD analysis of the cooling system [67].

Figure 4.8   High torque density BPM for electric racing cars [69].

The use of perfluorinated inert liquids are recommended. However, these fluids have properties that are strongly dependent on the working temperature [75,77].

Two liquids are considered for the submerged double-heat impingement (SDJI) method, both are Baysilone silicon oils manufactured by BAYER [77].

An important selection criteria for cooling fluids is the minimization of the external case-to-fluid thermal resistance. A fluid thermal figure-of-merit (FOM) is introduced. This figure has an empirical form and is strictly determined based on measurements. FOM can be associated with the heat transfer coefficients of the classical thermal analysis method [75].

For small ratio axial length/diameter of the machine, heat transfer predominantly occurs in the impingement region, represented by the following figure-of-merit:

$$FOM_j = \frac{d^{0.5} k^{0.6} c^{0.4}}{\mu^{0.1}} \tag{4.21}$$

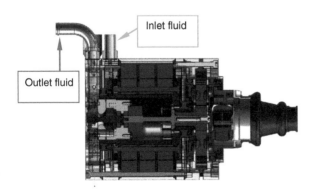

Figure 4.9   Axial cross-section for high torque density motor with dual cooling system: oil through the slots and air through the rotor [69]. *Source*: Reproduced with permission of IEEE.

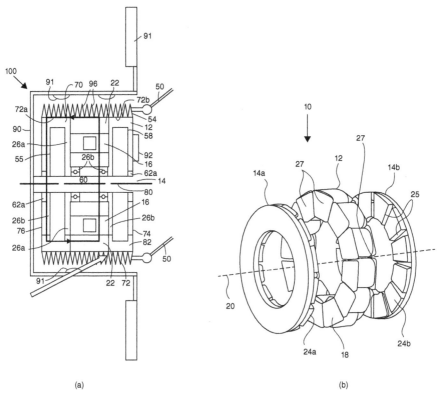

(a)   (b)

Figure 4.10   YASA motor cooling system (a) axial cross-section, (b) 3D view of the motor—US 20130187492 A1 [71].

For large ratio axial length/diameter, the wall jet region dominates the heat transfer:

$$FOM_{wj} = \frac{d^{0.8} k^{0.6} c^{0.4}}{\mu^{0.4}}$$

where d = density (kg/m^3); k = thermal conductivity (W/mK); c = specific heat capacity (J/kg°C); μ = dynamic viscosity (kg/m^{2}s).

Table 4.5 shows that the heat transfer coefficients of the impingement area are not influenced by the fluid properties [75, 77, 78]. Only the area where the fluids are in direct contact (wall jet) will be characterized by heat transfer coefficients that vary with the cooling fluid properties. The optimal cooling will also depend on the geometry of the nozzles.

Figure 4.11 shows a 3D schematic representation of a brushless permanent motor with spray cooling system. The fluid flow directions coming from various nozzles mounted on the shaft, rotor poles, or end winding to outer cap and stator bore are modeled via symbolic arrows.

TABLE 4.5 Physical properties of fluorinert liquids (3M) and Baysilone oils (BAYER)

Property (at 25 °C)	DI-water	Fluorinert oils						Baysilone oils		
		FC-72	FC-84	FC-77	FC-40	FC-43	FC-70	HY-M3	H5S-M5	H-M50
Boiling point (°C)	100	56	80	97	155	174	215	65	120	250
Pour point (°C)	0	−90	−95	−95	−57	−50	−25	−60	−40	20
Density (kg/m^3)	998	1680	1730	1780	1870	1880	1940	900	920	960
Thermal conductivity (W/m°C)	0.598	0.057	0.060	0.063	0.066	0.066	0.070	0.105	0.116	0.150
Specific heat capacity (J/kg°C)	4180	1046	1046	1046	1046	1046	1046	1510	1510	1510
Dynamic viscosity (kg/m^{2}s)	10×10^{-4}	6.7×10^{-4}	9.5×10^{-4}	14×10^{-4}	41×10^{-4}	53×10^{-4}	272×10^{-4}	27.6×10^{-4}	46×10^{-4}	480×10^{-4}
Dielectric constant at 1 kHz	78	1.76	1.81	1.86	1.89	1.90	1.98	2.5	2.5	2.8
Volume resistivity (ohm ×cm)	18×106	1×1015	1×1015	1.9×1015	4×1015	3.4×1015	2.3×1015	1×1014	1×1014	1×1015
FOM$_i$	1300	246	249	250	237	231	207	261	267	251
FOM$_{wj}$	82000	20472	18795	16955	11800	10694	5906	11785	10380	4904

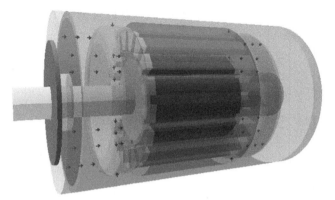

Figure 4.11 Spray cooled motor.

In Figure 4.12, another combined cooling system with stator water jacket, oil as cooling fluid, and forced air flow through the rotor assembly is presented.

As illustrated in Figure 4.13, the motor/generator may be in a cooling circuit with the transmission such that the cooling oil is provided from the transmission to the motor/generator, that is, the same cooling oil used to cool transmission components cools the motor/generator. Alternatively, the motor/generator may be provided with cooling oil in a separate cooling circuit not shared with the transmission. A power inverter module operatively connected with the motor/generator for providing electrical power to the stator may be cooled within the same cooling circuit as the motor/generator, or may have a dedicated cooling circuit using oil, water–ethylene glycol, or air cooling.

Figure 4.12 Nissan leaf traction motor—water jacket with oil and air induction system [70].

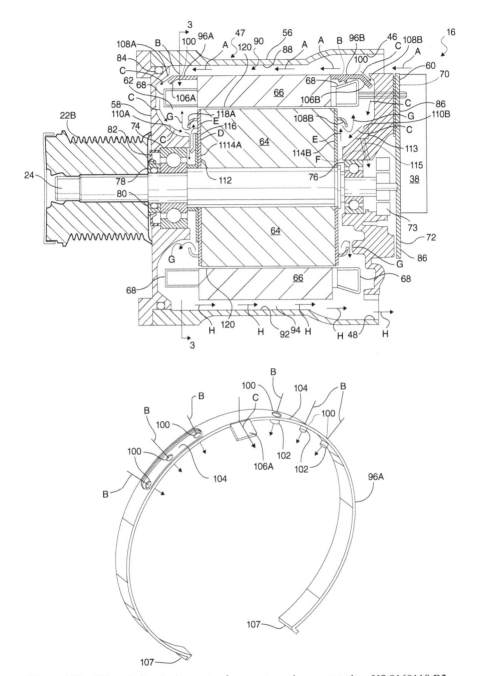

Figure 4.13   Oil cooled motor/generator for an automotive powertrain—US 8169110 B2 [72].

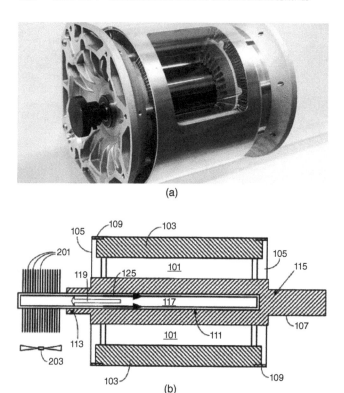

Figure 4.14   (a) Tesla S induction motor, (b) co-axially cooled stator and rotor—US 2014/0368064 A1 [73].

In this solution, only one cooling path is used to deliver cooling fluid to cool the stator, the rotor, and preferably bearings within the motor/generator. The flow is controlled using a ring so that the cooling fluid is thrown outward toward the stator by centrifugal force.

In Figure 4.14, the cooling system of the electric motor contains one heat pipe that goes through the motor hollow axle. An end of the heat pipe is extended and coupled to a heat exchanger that is formed by metallic plates shaped as fan blades.

4.4   HEAT EXTRACTION THROUGH RADIATION

Radiation heat transfer mode from a surface appears due to the energy transfer by electromagnetic waves [1–4]. The energy is emitted by vibrating electrons in molecules of the material of the surface of the analyzed body. The amount of emitted energy depends upon the absolute temperature of the body and can occur also in vacuum environment.

TABLE 4.6   Total emissivity of various materials

Material	Emissivity	Material	Emissivity
Aluminum		Iron	
Commercial sheet	0.09	Oxidized	0.74–0.89
Unoxidized	0.03–0.1	Unoxidized	0.05
Heavily oxidized	0.20–0.30	Nickel oxidized	0.31–0.46
Alloy A3003, oxidized	0.41	Paints	
Alumina on Inconel	0.45–0.69	Black	0.92
Asbestos	0.96	White	0.91
Carbon	0.79–0.95	Green	0.95
Chromium polished	0.06	Steel cold rolled	0.75–0.85
Copper		Rubber, hard	0.94
Polished	0.03	Silver, polished	0.02–0.03
Black oxidized	0.78	Inconel polished	0.19–0.21
Brass polished	0.03	Tin unoxidized	0.04

Radiation depends on emissivity ε and the view factor F of the analyzed surface, the ambient temperature T_0, and housing body temperature T_1, with the corresponding thermal resistance given by:

$$R_{th} = \frac{(T_1 - T_0)}{\sigma \varepsilon F \left(T_1^{\,4} - T_0^{\,4}\right) A}$$

Radiation phenomena occur both inside and outside the motor and in parallel with heat transfer through conduction and convection, natural or forced. It is possible to improve the motor cooling system through radiation by one or more of the following ways:

- Housing emissivity. The motor paint and its fin dimensions can be chosen to improve the motor apparent emissivity [8], Table 4.6.
- Reduce the temperature of the surroundings bodies. Hot devices, such as other motors working in the neighborhood, can transfer heat by convection, conduction, and radiation. For example, by means of a separator, the temperature of surrounding elements will not significantly influence the heat extraction through radiation for the analyzed component.
- Environment system absorptivity. When the motor is in a small working environment, for example, in a small box, a high wall absorptivity reduces the reflected radiation heat that strikes the motor.

Significant radiation thermal exchanges are not only present on the external frame, but also at several surfaces inside the motor. In particular, the main radiation paths are between the copper wires inside the slot and the stator lamination and between the end winding and the external frame.

4.5 COOLING SYSTEMS SUMMARY

Considering all the heat extraction methods described in previous sections, one can note that the forced convection method is the most efficient way to cool electrical machines. The clear majority of power traction electrical machines are cooled using forced liquid cooling. However, these cooling systems are also the most expensive and represent significant manufacturing challenges. The fluids must be well contained in sealed housing. The water jacket type housings add to the overall weight of the motor-drive system.

Natural convection and thermal conduction cooling represent the inherent methods to extract heat and therefore are also the cheapest from the manufacturing point of view.

Cooling through radiation is less effective, even though the cost of painting the motors represents generally a small fraction of the total production cost.

4.6 THERMAL NETWORK BASED ON LUMPED PARAMETERS

The thermal network analysis can be sub-divided into two main calculation types: heat transfer and flow network analysis.

a. Heat transfer analysis is the thermal counterpart to electrical network analysis with the following equivalences: temperature to voltage, power to current, and thermal resistance to electrical resistance.

b. Flow network analysis is the fluid mechanics counterpart to electrical network analysis with the following equivalences: pressure to voltage, volume flow rate to current, and flow resistance to electrical resistance. In the heat transfer network, a thermal resistance circuit describes the main paths for power flow enabling the temperatures of the main components within the machine to be predicted for a given loss distribution. In the thermal network model, it is possible to lump together components that have similar temperatures and represent each as a single node in the network. Nodes are separated by thermal resistances that represent the heat transfer between components. Inside the machine, a set of conduction thermal resistances represents the main heat transfer paths, such as from the winding copper to stator tooth and back iron, from the tooth and stator back iron nodes to the stator bore and housing interface In addition, internal convection and radiation resistances are used for heat transfer across the airgap and from the end windings to the endcaps and housing. External convection and radiation resistances are used for heat transfer from the outside of the machine to ambient.

Lumped circuit thermal models have been extensively utilized and validated on numerous machine types and operating points. Such a wide range of studies has increased confidence in such thermal models. As an example of this approach, the thermal model in Figure 4.15 has been used to analyze a 22.5 kVA synchronous machine, shown in Figure 4.16.

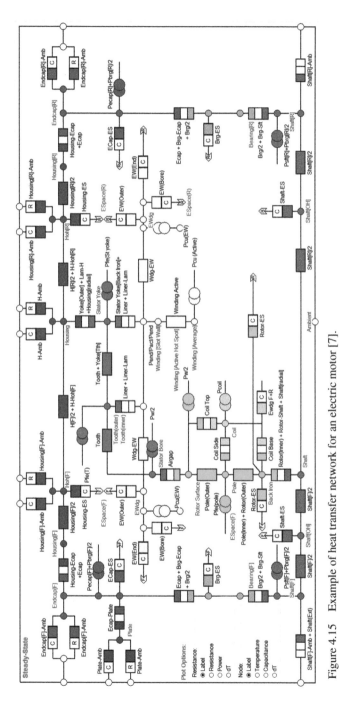

Figure 4.15 Example of heat transfer network for an electric motor [7].

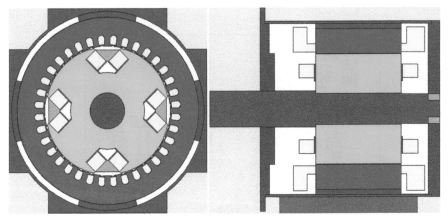

Figure 4.16   Radial and axial cross-sections of a synchronous wound field machine [7].

The model calculates both the air flow and heat transfer in the machine. Air flow and temperature rise for all stator and rotor nodes were within 10% of measured values.

Analytical lumped circuit techniques are also very useful in determining the thermal model's required discretization level. This refers to the number of sections used to model the electrical machine as a whole, or some of the more critical components, both in the axial and radial directions. Several studies have been performed to determine the required discretization level for a synchronous generator, with particular attention being given to the winding area. Due to its low thermal conductivity (2–3 W/m/K), this area is of great thermal significance and has to be analyzed with care. In the real winding, the heat generation is distributed over the section, and this study highlights the impact upon accuracy of specifying such loss in discrete nodes. A number of rotor winding models were used, ranging from a "single block" (1 × 1) representation to a rotor winding represented by 100 smaller sections (10 × 10). These two models are shown in Figure 4.17.

In Figure 4.18, the trend of the predicted averaged node temperatures as a function of the number of network nodes per section is reported. Concentrating all loss in one node in 1 × 1 network results in an unrealistic gradient between wall and winding center. Thus a suitable formula must be used to derive the average section temperature (20.2°C) from a single node temperature and wall temperatures; otherwise it could be wrongly interpreted as the whole winding section being at 60.5°C. The winding discretization level of 10 × 10 yields a more accurate prediction (average 17.7°C, peak 35.8°C) without the need for the formula when compared with FEA result (average 17.0°C, peak 37.2°C). To sum up, using lower levels of discretization reduces the accuracy of the results, whilst increasing node number unnecessarily complicates the thermal model.

As previously reported, lumped thermal parameters analysis involves the determination of thermal resistances.

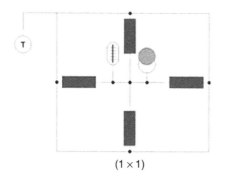

(1 × 1)

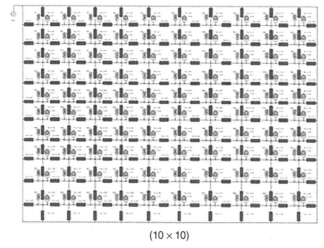

(10 × 10)

Figure 4.17   1 ×1 (top) and 10 × 10 (bottom) rotor winding models [7].

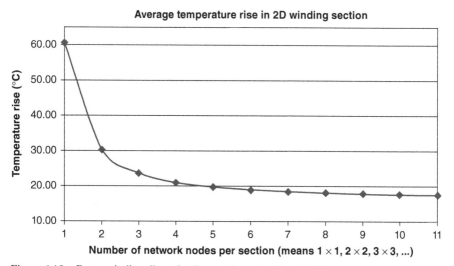

Figure 4.18   Rotor winding discretization results up to 11 × 11 [7].

4.7 ANALYTICAL THERMAL NETWORK ANALYSIS

The main characteristics of analytical software packages used in thermal analysis of electrical machines are discussed in this section. Analytical thermal network analysis software packages can be sub-divided into three types; dedicated software for thermal analysis of electric motors, general purpose network solvers with library components that can be used for thermal analysis, and custom packages written by the electrical machine designers for a particular machine type. There are very few commercial software packages for electric motor thermal analysis. Motor-CAD is one of the most widely used tools. The main advantage of the package is that the user need only input details of geometry, winding, and materials used and the software automatically sets up the thermal network and selects the most appropriate analytical formulations for each of the circuit components. Thus, the user need not be an expert in heat transfer analysis in order to use the software. The main limitation of such dedicated software is that the geometry is based on a fixed set of topologies, that is, pre-parameterized models. If the user's geometry is not similar to any of the inbuilt topologies, then a model cannot be implemented. Ideally, such packages should have some advanced features such that the user can edit the thermal circuit to model minor modifications to the geometry. It is important to underline that most motor topologies are relatively standard from the thermal point of view.

If a totally new type of structure is being analyzed, then there can be advantages in the user developing a completely new thermal network in a system simulation tool. Such packages usually have powerful network editors with drag and drop interfaces. This tool has features to help the user set up thermal networks in terms of thermal libraries. In particular, powerful "wizard" dialogs are used to select geometry and material data from which the most appropriate analytical mathematical formulations are automatically selected. Thus, the user sets up the thermal circuit from a geometric point of view rather than spending time researching heat transfer formulations. This is also the typical approach used by a motor designer when developing thermal analysis software for their own company use. As an example, for established machine topologies for which design evolves slowly, manufacturers may be inclined to adopt their own software, where impressive user interfaces and topology flexibility are replaced with available measured results, allowing validation and fine tuning of such software. One example worth mentioning is TTOOL, an analysis program for the design of synchronous generators ranging from 7.5 to 2500 kVA. The software development was carried out in accordance with Design for Six Sigma practices and the process took approximately 1 year. The appropriate discretization for the geometry and depth of analysis of each involved physics discipline was defined, so that on the basis of 90 machine input parameters (stored in a database for all frame sizes and core lengths) the resulting predictions typically lie within 5% of measured average winding temperatures. Simultaneous iterative solutions of heat generation, fan and flow circuit parameters, surface heat transfer, and conduction in solid material can be carried out for both steady-state and transient simulation. In addition to geometry and material changes, such effects as varied load, ambient conditions, filter or blockage effects can be studied with the package.

A further advantage of commercial packages used for thermal analysis is that they can be programmed to use sophisticated integration techniques that are tolerant to stiff sets of equations and nonlinearity. Stiffness can be a major problem when calculating the thermal transient response of motors which are constructed with parts having very different mass values.

For example, as the air-gap thermal capacity is much less than that of the winding, it will influence the integration step length, but have little influence on the thermal response of the machine. Also, network-based solvers that represent the system in terms of differential algebraic equations rather than ordinary differential equations have advantages in terms of stability for very nonlinear systems.

It is important to highlight that to obtain an accurate thermal model for an electrical machine, both analytical formulation and numerical method benefit from the previous experience of the designer. This is due to some thermal phenomena being dependent upon the manufacturing process of components.

4.8 THERMAL ANALYSIS USING FINITE ELEMENT METHOD

Finite element analysis (FEA) is now a standard tool for electromagnetic analysis. Both 2D and 3D models are used. Often software packages for electromagnetic analysis also include a module for thermal analysis. At first glance, FEA seems more accurate than thermal network analysis. However, it does suffer from the same problems previously described with uncertainty in the computation of thermal resistances due to interfaces and convection. In fact, an accurate FEA solution requires the knowledge of the same thermal parameters discussed in Chapter 4.2. A superficial knowledge of the geometrical and material properties used in a machine construction is often not sufficient to give an accurate prediction of the thermal performances. The main role of FEA is in the accurate calculation of conduction heat transfer in complex geometric shapes such as heat transfer through strands of copper in a slot. For this problem, FEA analysis can be used to calculate the equivalent thermal conductivity that can then be used in network analysis.

One problem with this calculation is that some assumptions must be made regarding the randomness of the conductor placement, the impregnation goodness, and any gaps between the slot liner and the stator lamination. This approach is much easier for winding types that have a known conductor placement, that is, form wound windings or precision windings.

In Figure 4.19, a steady-state thermal 2D FEA solution of a set of rectangular shaped copper conductors in a slot is shown. A fixed temperature boundary condition is applied to the outer surface of the stator lamination and a fixed amount of copper loss is applied to the problem. This simple boundary condition is possible as the designer is only interested in calculating the temperature difference between the winding hotspot and tooth/stator back iron. It is important to underline that simple thermal resistances in the lumped circuit can then be calibrated and used to give the

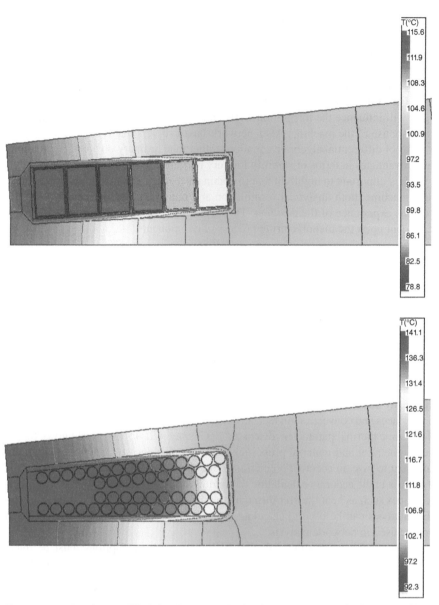

Figure 4.19   Steady-state 2D finite element analysis of the temperature rise in a slot with rectangular copper conductors.

same temperature rise, avoiding time-consuming tasks such as mesh definition and heat field computations.

FEA can be used to modify thermal networks to take into account specific loss distribution. In many lumped parameter thermal networks, it is assumed that the loss distribution across the electrical machine is uniform, but FEA results clearly illustrate

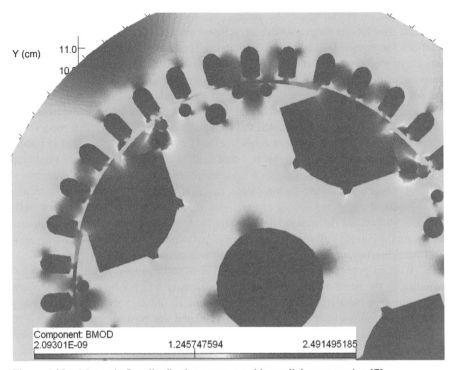

Figure 4.20   Magnetic flux distribution across machine radial cross-section [7].

the non-symmetrical nature of the operational flux density and related power loss distribution inside the machine as shown in Figure 4.20 for a salient pole synchronous machine.

Therefore, FEA results can be used to define the thermal network discretization level and a more realistic injection of the losses in the network nodes. It is important to underline that FEA applications are very time consuming for the actual geometry discretization and modification, even if a parametric approach for the geometry definition is used. Consequently, as most FEA electromagnetic packages include finite element thermal analysis facilities, FEA can be considered a convenient solution in very complex geometry not approachable with lumped parameters.

4.9   THERMAL ANALYSIS USING COMPUTATIONAL FLUID DYNAMICS

CFD applied to the design of an electrical machine primarily aims to determine coolant flow rate, velocity, and pressure distribution in the cooling passages or around the machine and the levels of surface heat transfer for subsequent analysis of temperature in the active material and remaining solid structures. This method can replace the combination of traditional one-dimensional (1D) ventilation resistance networks based on correlations for pressure drops across local and friction resistances, and correlations for surface heat transfer coefficients. The history of CFD use for studying aspects of electrical machines spans more than two decades back to days of simple purpose written CFD codes and early days of commercial codes. The limitations of the software and hardware of this era meant steep calculation costs and only little practical benefit for the industry. The industry benefited more by engaging in university research projects that evaluated the capability of CFD, as physical models for phenomena such as turbulence or rotation effects could lead to a variation in results [41]. Numerous papers were published that dealt with the comparison of predicted surface heat transfer coefficients with those measured experimentally or determined by established correlations. In the 1990s, the confidence in CFD and the evidence of its practical use in the design of machines started to emerge. Coolant flow optimization studies and isolated fan design were the most common examples. Computer hardware limitations still meant heavy use of periodic assumptions and coarse computational meshes, unless parallel computing was available.

Without proper understanding of fluid flow in or around the machines, continuation in the trend of increasing the power density will not be possible. Modern CFD codes are mostly based on the finite volume technique solving Navier–Stokes equations, complimented by a selection of validated and proven physical models to solve 3D laminar or turbulent flow and heat transfer to a high degree of accuracy. Major challenges with CFD vendors now lie in bringing the codes to a wider engineering community. Experts on CFD are not as much segregated from other engineers due to the in-depth knowledge of the underlying fundamental physics, but due to the skills it takes to convert geometry into a discretized mesh (or grid). The complexity of the meshing process lies in reducing the amount of detail in the machine without impacting on the accuracy of the solution.

In most cases, today, CFD analysis would start with some form of 3D native model produced in a CAD package. The deciding factor for choosing a CFD package is how tolerant it is to deficiencies in the particular geometry meshing software. Many CFD users, including those working with the leading commercial codes, have experienced periods of frustration when dealing with real geometry. When meshing internal volumes of electrical machines, one has to observe certain rules to avoid excessive numbers of cells in the mesh. Very narrow gaps between rotor and stator or in radial channels require high aspect ratio hexahedral elements, which usually rule out automated meshing techniques. Very good practice, often unavoidable, is sub-dividing the whole domain to volumes that are easier to mesh (Figure 4.21). The splitting is required by some CFD codes also to separate domains with rotation associated with rotating parts, and stationary parts.

To provide a rough guide on size of models and required computer power, a 64-bit workstation with 8 GB of RAM is a recommended minimum industry standard, typically needed to perform an analysis of air flow and convective heat transfer for a 180° periodic or full model, depending on level of details, for internal flow. This corresponds to discretization of the fluid region to approximately 8 million cells. In case of machines with many internal cooling ducts, 1-pole periodicity may result in

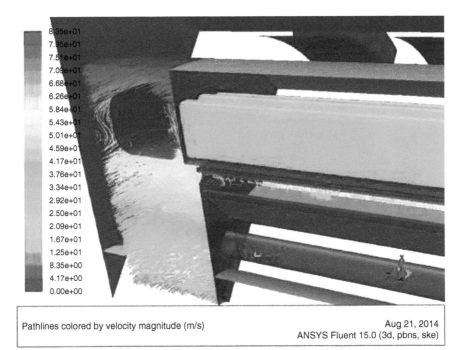

Pathlines colored by velocity magnitude (m/s)	Aug 21, 2014
	ANSYS Fluent 15.0 (3d, pbns, ske)

Figure 4.21   Model of the fluid flow around end winding in ANSYS Fluent.

consuming the same resource. The types of CFD analysis for an electrical machine can be divided into:

A. Internal flow—either in a through ventilated machine, where ventilation is driven by a fan (Figure 4.21) or self-pumping effect of rotor, or in a TEFC motor/generator to assess the air movements that exchange heat from winding overhangs to frame.

B. External flow—flow around the enclosure of a TEFC motor/generator (Figure 4.22).

C. Fan design and performance studies—due to cost of material, manufacturing processes, space or access constraints, fans employed in electrical machines often have very poor aerodynamic efficiency (Figure 4.22). In the case of radial fans, there are rarely any means of pressure recovery at exit. CFD offers a great deal of help in improving fan design and its interaction with the cooling circuit.

D. Supporting analysis—water flow in cooling jackets, cooling of associated power electronics (Figure 4.23).

The secondary function of CFD can be to solve heat flow paths all the way into the regions of their origin by means of conduction. This is essentially extending the CFD

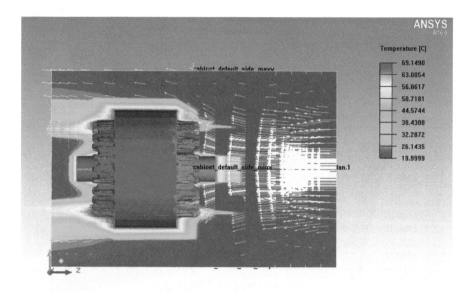

Figure 4.22   Model of the fluid flow around stator pack in ANSYS Icepak.

capability by FE thermal solution, often referred to as conjugate heat transfer modeling. In a R&D environment, it may be useful in prototype design work for validating lumped parameter thermal models, otherwise, in most cases it devalues the primary objective of CFD by introducing additional assumptions related to manufacturing processes and material properties. A link between CFD and a lumped thermal parameter circuit is much more meaningful for a design engineer wishing to perform analysis of a number of design iterations and settings. It has been proven in the design office environment that including solid regions in a CFD analysis prolonged preparation by increasing the complexity of setup, with little appreciable benefit. On the other hand, the conduction modeling capability of CFD can be used separately by

Figure 4.23   Fluid flow and temperature gradients in a three-parallel channels housing jacket (ANSYS Fluent).

disabling flow equations (leaving just one equation for energy) to undertake studies such as detailed temperature distribution in stator slots.

While attempting the CFD analysis for an electrical machine, the most common assumptions or simplifications are:

- Periodicity—depending on the position of inlets and outlets, terminal connections, etc. in the case of internal flow, or orientation in the environment for external flow, this may impact upon the accuracy of results and requires good judgment on the part of the analyst.
- Steadiness of flow and heat transfer with rotation—transient flow solutions are in most cases unnecessary, as time-averaging models give close results at a fraction of time. Some heat transfer augmentation due to pressure waves from rotor onto stator and back in salient pole machines was reported, however such differences are small and local.
- Surface roughness—data for surface roughness are very difficult to obtain and particularly difficult to input, as many surfaces have a specific type of roughness, such as one-directional roughness, or some surfaces are more uneven than rough.
- Geometry of complex structures can be simplified, for example, porosity model can be applied to simplify the geometry of bar-wound overhangs.

The following most common mistakes are made when attempting a CFD analysis:

- Not allocating sufficient computer resource, resulting in too coarse a computational mesh.
- Deficiencies in computational mesh (highly skewed cells)—can vary in effect from inability to obtain a solution to large local errors in flow field and heat transfer.
- Unsuitable type of cells resulting in excessive mesh size, often due to automated meshing, for example, using tetrahedral cells in rotor–stator airgap instead of high aspect ratio hexahedral cells.
- Poor definition of boundary conditions—results of any analysis can be only as good as the input data.

There are multiple ways of looking at the role of CFD in the future. There is no doubt that CFD as a discipline will become more popular and widespread, a cost-effective tool for innovation. The vendors and academic circles will continue to enhance the capability of commercial codes for further more complex physics phenomena, this is, however, unlikely to influence accuracy of solutions for electrical machines as achieved today. From the industrial viewpoint, CFD can be used to identify possibilities for segmentation of product ranges, for example, for high efficiency continuous operation or low-cost stand-by units. It is within the scope of CFD to evaluate efficiency improvements in electric machines that can be translated to cost savings during operation.

4.10 THERMAL PARAMETERS DETERMINATION

As discussed in the previous sections, the accuracy of both sophisticated and simple thermal networks is dependent upon several parameters for which reliable data may be difficult to find. In fact, many of the complex thermal phenomena inside electric machines cannot be solved by pure mathematical approaches using a closed relationship. Designers with extensive working experience on similar designs using comparable manufacturing processes can make a correct choice of such parameter values. For designers approaching their first thermal analysis, these choices are more difficult. It is particularly important for the user to have available reasonable starting values for the less well-known and defined parameters. It is essential to use reliable relationships in the determination of the more complex thermal parameters. The weight of these parameters on the thermal analysis results has been investigated from the sensitivity point of view. Such information can be used successfully to set default parameters in thermal analysis software to give reasonably accurate predictions at the start of the design process before manufacturing methods and tolerances have been fully thought-out. Expected upper and lower limits of such parameters based on experience can be built into automated sensitivity analysis so that the designer can quickly access the main constraints to cooling and quantify the effects of manufacturing options and tolerances.

In the following, a summary of the obtained results are reported.

4.10.1 Equivalent Thermal Resistance Between External Frame and Ambient Due to Natural Convection

In totally enclosed machines with no fan or a shaft-mounted fan operating at slow speed, the thermal resistance R_0 (°C/W) between the housing and ambient is often the largest single resistance between winding and ambient. When the total area A of the external frame is known, (4.22) can be initially used.

$$R_0 = 0.167 A^{1.039} \qquad (4.22)$$

An alternative is to make a first estimate of the convection and radiation heat transfer coefficients and calculate R_0 using equations (4.9)–(4.11). Typically, the combined natural convection and radiation heat transfer coefficient lies in the range 12–14 W/(m^2 °C) for simple geometric shapes [49].

4.10.2 Equivalent Thermal Conductivity Between Winding and Lamination

It is widely recognized that the thermal behavior of the wires inside the slot is very complex. The thermal resistance can be computed using (4.1), but the value of the thermal conductivity k is not easily defined. A possible approach to simplify the thermal resistance computation is to use an equivalent thermal conductivity of the system winding impregnation and insulation ($k_{cu,ir}$). This equivalent thermal conductivity depends on several factors such as material and quality of the impregnation, residual air quantity after the impregnation process. If the equivalent thermal conductivity $k_{cu,ir}$ is known, the thermal resistance between the winding and the stator lamination can be easily computed. When the slot fill factor k_f, the slot area A_{slot}, and axial core length L_{core} are known, (4.23) can be used as a rough guide.

$$k_{cu,ir} = 0.2749[(1 - k_f)A_{slot}L_{core}]^{-0.4471} \qquad (4.23)$$

The quantity inside the square bracket represents the available net volume for the wire/slot insulation and the impregnation inside the slot.

An alternative approach is to sub-divide the winding in the slot into a number of thermal resistances from the slot center to slot wall. The resistance values can be calculated from knowledge of the slot shape, slot fill, and impregnation goodness.

4.10.3 Forced Convection Heat Transfer Coefficient Between End Winding and Endcaps

The thermal resistance between winding and endcaps due to forced convection can be evaluated by (4.5). Again the value of h_c is not simple to define. The value of h_c can be evaluated by (4.24) as a function of the air speed inside the motor endcaps,

$$h_C = 6.22\,v \qquad (4.24)$$

or by (4.25) to account for combined natural and forced convection.

$$h_c = 41.4 + 6.22\,v \qquad (4.25)$$

Other alternative relationships are also available [24, 29, 45, 49] and give similar results.

4.10.4 Radiation Heat Transfer Coefficients

The thermal resistance for radiation can be evaluated using (4.5) and (4.23), when h_R is available. Inside and outside the motor, several parts exchange heat by radiation. In some cases, such as aerospace applications, all the heat transfer is due to radiation. The following values of the radiation heat transfer coefficients can be initially used:

8.5 W/(m^2 °C)   between copper and iron lamination;
6.9 W/(m^2 °C)   between end winding and external cage;
5.7 W/(m^2 °C)   between external cage and ambient.

4.10.5 Interface Gap Between Lamination and External Frame

The interface gap between the lamination and the external frame is due to imperfections in the touching surfaces and it is a complex function of material hardness, interface pressure, smoothness of the surfaces, and air pressure. The interface gap between stator lamination and external frame is very important because most of the motor losses cross this surface. For industrial induction motors, interface gap values between 0.01 mm and 0.08 mm have been found. As the interface gap between the lamination and the external frame is not only dependent on the frame material, but is strongly influenced by the stator core assembly and by the core-external frame inserting process, it is not possible to compute its value. A value of 0.03 mm can be considered a reasonable value to be used as a default at the start of the design process. Sensitivity analysis with the interface gap varied between 0.01 mm and 0.08 mm.

4.11 LOSSES IN BRUSHLESS PERMANENT MAGNET MACHINES

4.11.1 Introduction

The BPM machine represents the electrical machinery topology with the highest torque density [2]. In the last three decades, BPMs have seen very significant development and growth of manufacture from various industrial applications; hybrid and electric vehicles, renewable energy generation, aerospace, and home appliances are a few of the applications. The growth is being driven by the rare-earth element's extraction and processing, with high energy magnets used widely in the BPM manufacturing. Theoretically seen as an everlasting source of energy within the electrical machine system, permanent magnet materials may be irreversibly demagnetized and hence lose its excitation energy due to the thermal stress and high-fault electrical loads.

There are two main high energy magnetic materials that are currently employed: Neodymium Iron Boron (NdFeB) and Samarium Cobalt (SmCo) and in high power motors they tend to be used in their sintered form. Figure 4.24 shows a typical set of demagnetization curves for an NdFeB magnet. When the magnet operation point is below the knee point of the demagnetization curve, the magnet is irreversibly demagnetized. It can be seen that this is highly dependent on the operating temperature.

The thermal stress on permanent magnets is created by the losses dissipated in the machine which will heat the magnets and need to be dissipated. One can thermally protect the permanent magnets by reducing the local losses, that is, the induced eddy current losses in the magnet blocks, and/or using an efficient cooling system. Depending on the application, cooling systems can be employed with natural convection (TENV), forced convection (air or liquid cooling), or radiation cooling (in the case of BPMs, operating in vacuum environment).

The thermal analysis of an electric motor is generally regarded as more challenging compared to the electromagnetic analysis in terms of the ease of construction of a model and achieving good accuracy. This is because it is highly dependent not only on the design but also on the manufacturing tolerances.

The thermal analysis of a BPM motor is a 3D problem (again, electromagnetic problems are often only 2D with additional end effects) which requires complex heat transfer phenomena to be solved; for example, heat transfer through complex

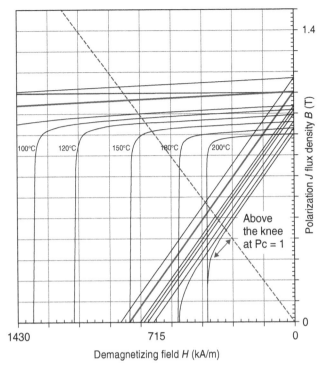

Figure 4.24   Demagnetization curves for NdFeB type magnet [1, 2].

composite components such as the wound slot, temperature drop across interfaces between components, and complex turbulent air flow within the endcaps. The latter requires analysis around the end winding that includes rotational effects.

This study presents various solutions for an efficient thermal management of the permanent magnets in BPMs. The mechanism for the main loss components in BPMs is briefly described: stator copper losses, iron losses, and magnet losses.

4.11.2   Stator Copper Losses

Generally, the main loss component in a BPM is the stator winding copper loss. These losses are a function of current and stator winding resistance. In BPMs, the temperature affects both the required current for the motor to deliver an imposed output torque and the electrical resistivity of the material used to build the stator winding.

An increase in winding temperature gives an increase in copper resistivity according to the formula:

$$\rho = \rho_{20} \cdot [1 + \alpha \, (T - 20)] \tag{4.26}$$

where $\rho_{20} = 1.724 \times 10^{-8}$ Ωm and $\alpha = 0.00393/°C$

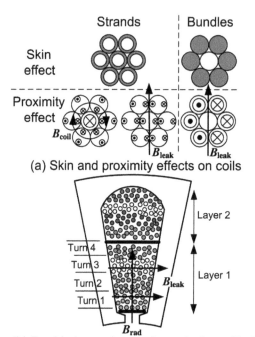

(a) Skin and proximity effects on coils

(b) Double layer slot random winding with 4
turns or coil and 22 strands-in-hand

Figure 4.25   Skin and proximity effects in a bundles coils and strands-in-hand random wound coils in slot.

Thus, a 50°C rise gives 20% increase in resistance and a 140°C rise gives 55% increase in resistance. A more difficult phenomenon to estimate is the increased stator winding resistance due to the frequency or AC losses; that is, the skin effect and the proximity effect. BPMs are often wound with coils spanning only one tooth. In high speed applications, parallel paths may be needed. For a high current machine, these parallel paths may be within the coil itself and often referred to as "strands-in-hand" or multiple bundle strands. However, parallel paths for coil turns located in the slot top and slot bottom will experience a proximity effect since there will be some flux linkage due to leakage. This should be differentiated from skin effect in windings. The differences are noted and Figure 4.25 illustrates these. Even with small wire gage like the Litz wire configuration used to eliminate skin effect, proximity effects will exist.

The strands-in-hand and parallel paths are used in BPMs to mitigate the proximity losses. Typical results are shown in Figure 4.26. This shows the instantaneous loss due to the skin and proximity effects. The example machine has the stator winding with nine turns per coil connected in series and eight parallel strands. The conductors are rectangular 0.5×1.0 mm. Figure 4.27 shows the instantaneous current distribution that was obtained from a 2D finite element analysis at 2000 Hz. The coil side that is just at the leading edge of the rotor pole has a considerable amount of variable magnetic flux leaking through that region. The maximum current density $= 260$ A/mm^2.

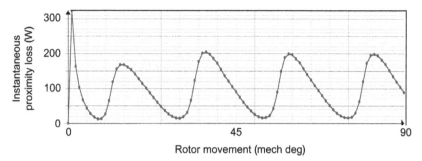

Figure 4.26   Instantaneous AC losses per slot variation with rotor position at high speed.

The ratio of AC to DC resistance is 10; the P_{DC} (DC losses) = 270 W and P_{AC} (AC losses) = 2700 W. This illustrates that even with the use of strands-in-hand, there can be considerable problems with increased losses at high frequency. Figure 4.28 shows the ratio of P_{AC}/P_{DC} in the same slot for frequencies up to 2 kHz. Notice the massive increase in the AC losses for the conductors placed at the slot opening region.

Alternative solutions for mitigating the proximity losses have been proposed: twisted wires, winding arrangement with flat rectangular wire placed along the slot leakage-flux lines, and reduced slot fill factor with the copper wires pushed well within the slots and further away from the slot opening region.

4.11.3   Iron Losses

A successful electrical machine design optimization process requires the accurate prediction of iron losses. Power electronic converters, used to supply electric motors in variable speed systems, have a non-sinusoidal (PWM) voltage waveform that causes increased losses in the lamination steel. Estimation of the losses under these operating conditions represents a challenging task. For this purpose, a large number of models, which are based on a physical or an engineering approach, have been proposed by different authors and yet a definitive conclusion has not been reached. The

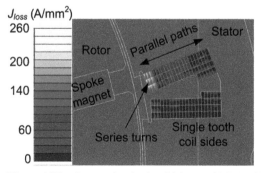

Figure 4.27   Current density in a high speed BPM with spoke magnets.

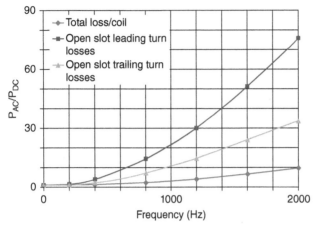

Figure 4.28   Ratio of the AC/DC losses with frequency for slot configuration given in Figure 4.27.

iron losses can be segregated in static (hysteresis) and dynamic (eddy current) losses. One good approach to calculate these specific losses is to use a function as follows:

$$w_{Fe} = w_{hys} + w_{eddy} = k_h\,(f, B) \cdot f B^2 + k_e\,(f, B) \cdot (f\,B)^2 \qquad (4.27)$$

where the loss coefficients k_h, k_e are determined using third-order polynomial functions.

The material properties are changing with magnetic load and frequency; the magnetic permeability μ decreases with frequency and induction as illustrated in Figure 4.29. This shows the case for a fully processed silicon steel M43. At higher frequency, the dynamic (eddy current) losses become the dominant iron loss component as displayed in Figure 4.30. Other factors leading to increased iron losses

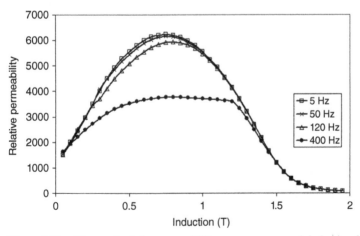

Figure 4.29   Measured relative permeability curves versus peak induction for different frequency levels—fully processed material M43.

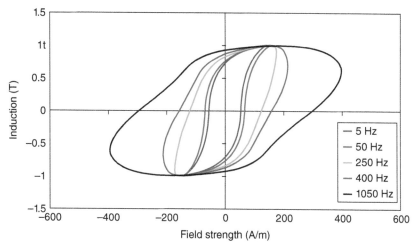

Figure 4.30   Dynamic hysteresis loops for peak induction 1 T—fully processed material M43.

in a BPM are the current time harmonics generated by the PWM inverters and the manufacturing process. In tested BPMs, the iron losses are expected to be at least double when compared to the ideal calculation cases based on catalogue data loss curves and assuming pure sinusoidal current waveforms.

In BPMs, the iron losses are created by the variation in the magnetic field due to the permanent magnet rotation and the pulsations in the stator winding magnetic field. The highest iron loss density in a BPM will occur in the stator laminated teeth region and on the rotor surface, see Figure 4.31. Notice that combined with the stator copper losses distribution, the largest amount of losses in BPMs is concentrated in the stator volume limited by the air-gap area.

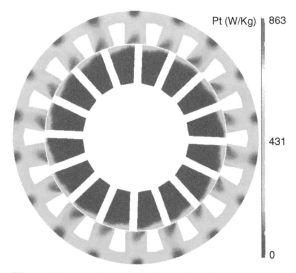

Figure 4.31   Iron losses distribution in a BPM under load conditions.

These losses will generate heat that has to be extracted or dissipated. In a BPM, the magnets are located either in the vicinity of the stator heat source, as in the surface-mounted PM rotor, or they can be better thermally protected by embedding them in the rotor lamination pack.

4.11.4   Magnet Losses

High energy permanent magnets (NdFeB and SmCo) are characterized by locally generated losses. These losses are created by the eddy currents induced by dips in the air-gap flux density due to slotting and by current time and space harmonics. The latter cause is more significant in a BPM with DC operation; that is, trapezoidal currents, when compared to the synchronous BPM, where the currents are sinusoidal and have a very low THD. Depending on the number of slots/pole and the winding configuration, the magnet losses may be significant even for low speed applications. By comparison with low energy magnets such as ferrite, the rare-earth magnets have a much lower electrical resistivity. NdFeB and SmCo magnets have an electrical resistivity approximately 100 times and 40 times, respectively, higher than copper. Table 4.7 summarizes the electrical resistivity values for some relevant materials used to manufacture electrical machines.

The variation of the electrical resistivity for sintered rare-earth permanent magnet materials is usually neglected, but recent studies show that Sm_2Co_{17} samples can have a linear increase of 6% in the electrical resistivity, that is, from 0.8 $\mu\Omega$m to 0.85 $\mu\Omega$m in transversal direction and from 0.87 $\mu\Omega$m to 0.90 $\mu\Omega$m in axial direction, when temperature increases from 50°C to 100°C. For the same temperature range, samples of sintered NdFeB magnets experience an increase of the electrical resistivity in axial direction from 1.6 $\mu\Omega$m to 1.65 $\mu\Omega$m and from 1.3 $\mu\Omega$m to 1.35 $\mu\Omega$m in transversal direction.

For example, in a spoke machine, the instantaneous magnet current at any one time is shown in Figure 4.32, while the average loss over one complete mechanical cycle is shown in Figure 4.33, illustrating that losses are concentrated at the top of the magnet.

In practice, when analyzing various magnets data, only in sintered NdFeB and SmCo materials is the locally induced eddy current loss such that it may be significant and require a technical solution to minimize them or to use a better cooling system for heat extraction.

TABLE 4.7   Electrical resistivity values (Ωm)

Material	Value
Iron	10×10^{-8}
Sm–Co 1-5 alloys	50×10^{-8}
Sm–Co 2-17 alloys	90×10^{-8}
NdFeB—sintered	160×10^{-8}
NdFeB—bonded	14000×10^{-8}
Ferrite	10^5

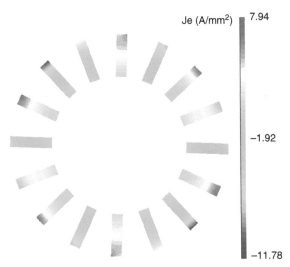

Figure 4.32   Distribution of the induced eddy current density in the magnets for one instance in time.

The mitigation of magnet losses is usually achieved using magnet segmentation along the magnetization lines in the magnet cross-section and/or axially.

If we consider m magnet blocks/pole in the x–y plane and n magnet blocks in the z direction, the ratio of the losses in a segmented configuration and the monolithic configuration in a motor with axial length L is approximated by the expression:

$$\frac{P_{\text{segmented}}}{P_{\text{monolithic}}} = \frac{L + \tau}{mL + n\tau} \tag{4.28}$$

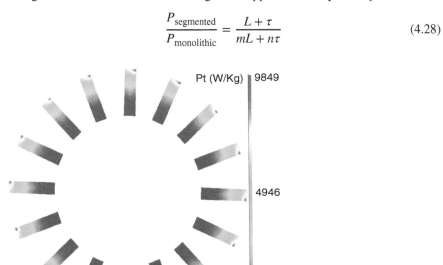

Figure 4.33   Distribution of specific magnet losses in a BPM under load conditions; losses are averaged for a full mechanical cycle.

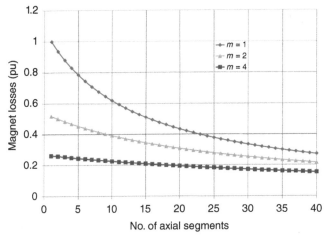

Figure 4.34   Effect of segmentation on the magnet losses.

Figure 4.34 shows the effect of axial segmentation on magnet losses, when there is one, two, or four magnet blocks per pole in x–y plane. Notice the higher impact on magnet losses minimization of the x–y plane segmentation.

In BPMs with surface-mounted magnet rotor configuration, it is necessary to ensure the retaining of the magnet blocks through a special mechanical fixture that can withstand the mechanical stress created by the centrifugal force. Often retaining sleeves are employed. If the material of the retainer is electrically non-conducting, for example, fiber glass or with very high electrical resistivity, for example, composite carbon fiber, there are practically no extra losses on the rotor that could affect the magnet's temperature. However, non-metallic sleeves have a limited temperature operation at 180°C. For cost reduction reasons, there are solutions for retaining the magnets using non-magnetic metallic sleeves, for example, titanium alloys, stainless steel, brass, or aluminum [28].

4.12   COOLING SYSTEMS

The ultimate solution to ensure that an electrical machine operates at the required parameters is an efficient cooling system. All the loss components presented in Chapter 4.11 will generate heat that needs to be extracted without compromising irreversibly the magnet's properties. Typically for 100°C temperature rise, NdFeB magnets will lose 11% of the magnetic flux, SmCo magnets will lose 3% magnetic flux, and ferrite magnets will lose 20% magnetic flux.

Magnets are usually isolated from the main heat sources (stator) because they are located on the opposite armature. The thermal behavior of BPMs can be analyzed using lumped thermal circuit models. These models are similar to electrical networks and require a much lower computational effort than alternative numerical methods such as finite element (FEA) and CFD methods. Heat can be extracted through conduction, convection (natural and forced), and radiation.

Conduction heat transfer mode is created by the molecule vibration in a certain material. Typically, a material with good electrical conductivity is also characterized by a good thermal conductivity. In an electrical machine, it would be desirable to also have materials that are good electrical insulators and have good thermal conductivity. Metals, that is, copper, aluminum, steel, have high thermal conductivity due to their well-ordered crystalline structure. The thermal conductivity for metals, k, is usually in the range 15–400 W/m/K. Solid insulator materials, that is, resins, paper, do not have a well-ordered crystalline structure and are often porous. Thus, the energy transfer between molecules is impeded. The thermal conductivity for insulators is typically in the range 0.1–1 W/m/K. For comparison, air has an average thermal conductivity of 0.026 W/m/K.

With respect to the heat transfer via conduction, the most difficult region to model in an electrical machine is the stator slot area where the copper conductors are located together with the slot insulation, the conductor insulation, and the impregnation material. This problem is solved by using an equivalent thermal conductivity.

The effective properties of the two dominant isotropic materials in the slot can be used to estimate analytically the equivalent thermal conductivity in the slot using the formulations derived by Hashin or Milton. Hence, the equivalent thermal conductivity of the slot can be calculated using the thermal conductivities of the conductors and the slot impregnation. These are denoted as k_1 and k_2 respectively ($k_1 > k_2$). This approach assumes that the conductor insulation, that is, the enamel, has the same thermal conductivity as the impregnation material; that is, a resin type of material. If the conductors are randomly distributed within the slot, and represent f_1 volume ratio of the slot, while the impregnation occupies the volume ratio f_2 (where $f_1 + f_2 = 1$), we can define the equivalent thermal conductivity of the slot as:

$$k_e = k_2 \frac{(1 + f_1)k_1 + (1 - f_1)k_2}{(1 - f_1)k_1 + (1 + f_1)k_2} \tag{4.29}$$

Milton proposes two expressions, one for the lower limit of the equivalent property of the two-composite material (k_{eL}) and one for the upper limit of the equivalent property (k_{eH}) so that

$$k_{eL} = k_2 \frac{(k_1 + k_2)(k_1 + k_1 f_1 + k_2 f_2) - f_2 \zeta_1 (k_1 - k_2)^2}{(k_1 + k_2)(k_2 + k_1 f_2 + k_2 f_1) - f_2 \zeta_1 (k_1 - k_2)^2} \tag{4.30}$$

$$k_{eH} = k_1 \frac{(k_1 + k_2)(k_2 + k_1 f_1 + k_2 f_2) - f_1 \zeta_2 (k_1 - k_2)^2}{(k_1 + k_2)(k_1 + k_1 f_2 + k_2 f_1) - f_1 \zeta_2 (k_1 - k_2)^2} \tag{4.31}$$

where $\zeta_{1,2}$ represents a general material property.

For a slot with copper conductors ($k_{Cu} = 386$ W/m/K) and an impregnation resin with $k_{Resin} = 0.2$ W/m/K, the equivalent thermal conductivity of the slot, as calculated using (4.30) and (4.31), is plotted in Figure 4.35. The expression (4.31) for the upper limit of the equivalent property of the two-composite material was found to lead to substantially higher values and therefore not considered for the estimation of the equivalent thermal conductivity.

Convection heat transfer mode appears between a surface and a fluid due to intermingling of the fluid immediately adjacent to the surface, where a conduction

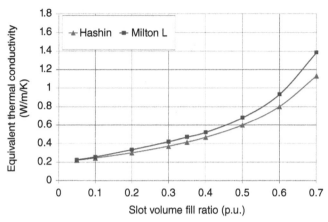

Figure 4.35   Slot equivalent thermal conductivity variation with slot volume fill ratio [26].

transfer mode will occur, with the remainder of the fluid due to the molecules motion. Generically, we distinct between: (a) natural convection when the fluid motion is due to buoyancy forces that arise from the change in density of the fluid in the vicinity of a surface; and (b) forced convection when the fluid motion is due to an external force created by a special device, for example, fans, pumps.

Based on the fluid flow type, it is possible to have laminar flow, that is a streamlined flow and occurs at lower velocity and turbulent flow that is created by the eddies at higher velocities when an enhanced heat transfer happens by comparison with the laminar flow case, but with a larger pressure drop.

Modern applications where BPMs are employed rely frequently on forced convection cooling systems that use air or liquid, that is, water, oil, and their combinations. Based on the convection technique, we can have: air natural convection ($h = 5$–10 W/m^2/K), air forced convection ($h = 10$–300 W/m^2/K), and liquid forced convection ($h = 50$–20000 W/m^2/K).

Forced convection can be achieved using configurations of channels, ducts, water jackets, spray cooling, and axle cooling:

- Natural convection (TENV) with various housing design types;
- Forced convection (TEFC) where the fin channel design is essential;
- Through ventilation with rotor and stator cooling ducts;
- Open end-shield cooling;
- Water jackets with various design types (axial and circumferential ducts) and stator and rotor water jackets;
- Submersible cooling;
- Wet rotor and wet stator cooling;
- Spray cooling;
- Direct conductor cooling; for example, slot water jacket;

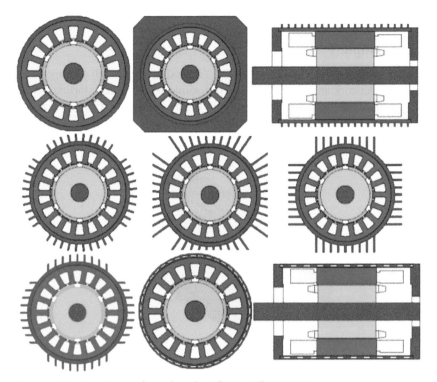

Figure 4.36   Housing configurations for BPMs cooling systems.

- Conduction where the internal conduction and the effects of mountings are relevant;
- Radiation, internal and external.

Figure 4.36 shows several housing configurations for cooling systems of BPMs.

The main difficulty in setting up an accurate thermal model for BPMs is in the circuit components that are influenced by manufacturing uncertainties. Commercial simulation software attempts to overcome these uncertainties:

- Air in the winding impregnation, how good a fit there is between the stator lamination and housing?, etc.;
- Measurements have been made on many machines in the past to set a set of default parameters for an average machine;
- Calibration using tests can be used to give better absolute accuracy;
- The process of calibration and comparing parameters with default values give an indication of how well the machine is constructed;
- Sensitivity analysis is recommended to gain an in-depth understanding of the main restrictions to cooling for a given design.

Several practical examples of cooling BPMs are given in the following section.

4.13 COOLING EXAMPLES

4.13.1 Example 1

A BPM for traction application (Figure 4.37a) is modeled both with a lumped thermal transient network and with FEA [2]. The estimated calculated temperature values are validated with experimental results in Figure 4.37b. The FEA analysis has useful features such as showing the temperature distribution throughout the solid

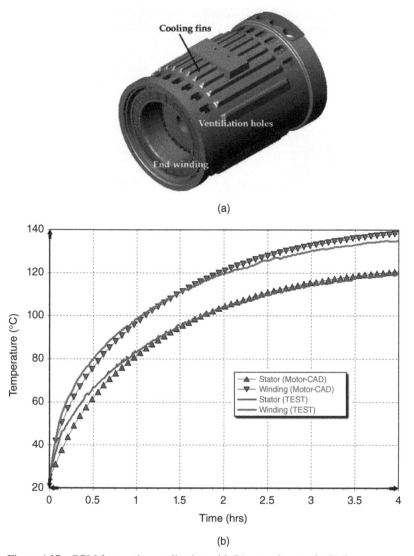

(a)

(b)

Figure 4.37   BPM for traction application with S1 operation results [14].

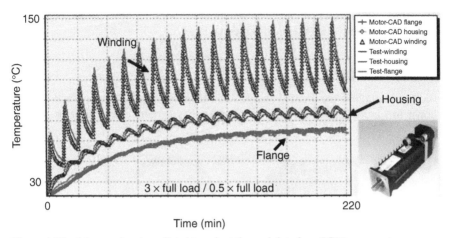

Figure 4.38   Measured and predicted transient thermal data for a BPM servomotor.

components. However, the FEA model shows no advantage in terms of accounting for the interface gaps and convection, radiation, and heat transfer through the composite set of components in the winding (copper + impregnation + air + slot liner, etc.).

4.13.2   Example 2

Thermal analysis on a BPM used as the brushless servomotor is shown in Figure 4.38. The lumped thermal network requires setting up with a small amount of additional data, that is, housing type, materials used for impregnation, housing, etc. The results show excellent correspondence between measured and calculated predictions of both steady-state and thermal transient data.

4.13.3   Example 3

Table 4.8 shows excellent agreement between analytical predictions and measured thermal transient data for the BPM servomotor shown in the inset. In this case, the duty cycle was a repetitive load of 1 minute with three times overload followed by

TABLE 4.8   Comparison between measured and simulated steady-state temperatures

Temperature measurement location	Measured (°C)	Simulated (°C)]	Error (%)
Winding hotspot	144.3	148.1	2.6
Housing	117.0	117.0	0.0
Tooth	127.9	129.4	1.0
End winding	134.1	126.8	−5.5
Endcap	107.4	107.9	0.5

10 minutes of half full load. It can be seen that the winding heats up and cools down at a much faster rate than the bulk of the machine.

4.13.4   Example 4

Another example is the thermal analysis on the outer rotor BPM axle traction motor shown in Figure 4.39. This type of machine has a particularly difficult path for

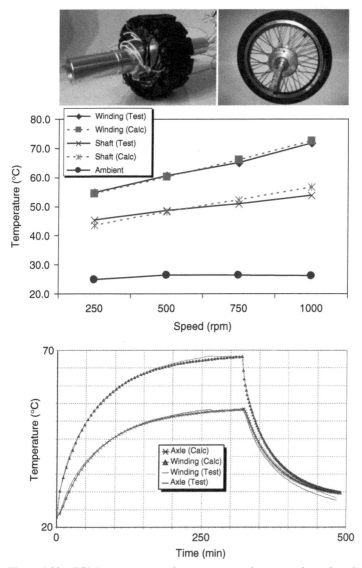

Figure 4.39   BPM outer rotor traction motor—steady-state and transient thermal results [18].

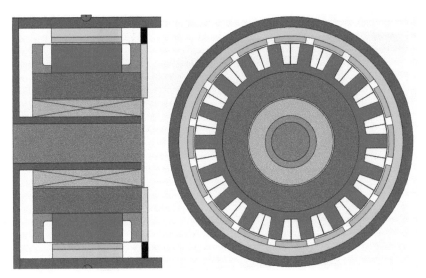

Figure 4.40   Axial and cross-section views for a BPM motor hub.

cooling of the stator winding as it is embedded inside the rotating outer magnets. All possible heat transfer paths are modeled using a lumped thermal network, that is, via the static axle to any heat sink attached, which is through the axle and the bearings to the rotating outer magnet retaining cylinder, or through the airgap and off the end winding to the rotating outer magnet retaining cylinder. Figure 4.16 shows a comparison between predicted and measured thermal steady-state and transient data. A satisfactory agreement is observed.

4.13.5   Example 5

Figure 4.40 shows the configuration of a BPM for a hub motor. The comparison between modeled and measured temperature values is given in Table 4.9. A continuous (steady-state) thermal map that considers all losses including AC losses in the stator winding and rotor losses (iron and magnet) is presented in Figure 4.41.

TABLE 4.9   Measured and calculated thermal values for BPM hub motor

Motor component	Open circuit (°C)		Short-circuit (°C)	
	Exp.	Sim.	Exp.	Sim.
Winding average	113	111	116	113
End winding NDE	105	108	112	111
End winding DE	107	110	114	113
Magnet	82	83	–	–

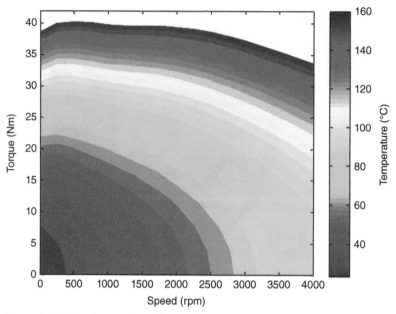

Figure 4.41   Continuous thermal map of the BPM motor hub.

REFERENCES

[1] M. Popescu, D. A. Staton, A. Boglietti, A. Cavagnino, D. Hawkins, and J. Goss, "Modern heat extraction systems for power traction machines—A review," *IEEE Trans. Ind. Appl.*, vol. 52, no. 3, pp. 2167–2175, May/Jun. 2016.

[2] M. Popescu, D. Staton, D. Dorrell, F. Marignetti, and D. Hawkins, "Study of the thermal aspects in brushless permanent magnet machines performance," in *IEEE WEMDCD'2013*, Mar. 11–12, 2013, pp. 60–69.

[3] R. Wrobel, P. H. Mellor, M. Popescu, and D. A. Staton, "Power loss analysis in thermal design of permanent-magnet machines—A review," *IEEE Trans. Ind. Appl.*, vol. 52, no. 2, pp. 1359–1368, Mar./Apr. 2016.

[4] Motor-CAD, www.motor-design.com

[5] http://en.wikipedia.org/wiki/Insulation_system

[6] http://www.ifam.fraunhofer.de/

[7] A. Boglietti, A. Cavagnino, M. Parvis, and A. Vallan, "Evaluation of radiation thermal resistances in industrial motors," *IEEE Trans. Ind. Appl.*, vol. 42, no. 3, pp. 688–693, May/Jun. 2006.

[8] G. I. Taylor, "Distribution of velocity and temperature between concentric cylinders," *Proc. R. Soc.*, vol. 159, pp. 546–578, 1935.

[9] A. F. Mills, Heat Transfer. Prentice Hall, 1999, ISBN 0-13-947624-5

[10] J. R. Simonson, Engineering Heat Transfer, 2nd ed. McMillan, 1998, ISBN 0-333-45999-7

[11] A. Bejan, Heat Transfer. John Wiley & Sons, 1993, ISBN 0-471-50290-1

[12] W. S. Janna, Engineering Heat Transfer, Van Nostrand Reinhold (International), 1988, ISBN 0-278-00051-7

[13] F. P. Incropera and D. P. De Witt, Introduction to Heat Transfer, John Wiley & Sons, 1990, ISBN 0-471-51728-3

[14] J. J. Nelson, G. Venkataramanan, and A. M. El-Refaie, "Fast thermal profiling of power semiconductor devices using Fourier techniques," *Trans. Ind. Electron.*, vol. 53, no. 2, pp. 521–529, Apr. 2006.

[15] Z. Gao, T. G. Habetler, R. G. Harley, and R. S. Colby, "A sensorless rotor temperature estimator for induction machines based on a current harmonic spectral estimation scheme," *IEEE Trans. Ind. Electron.*, vol. 55, no. 1, pp. 407–416, Jan. 2008.

[16] P. Mellor, D. Roberts, and D. Turner, "Lumped parameter thermal model for electrical machines of TEFC design," *IEE Proc.*, vol. 138, pp. 205–218, Sep. 1991.

[17] A. Boglietti, A. Cavagnino, M. Lazzari, and M. Pastorelli, "A simplified thermal model for variable speed self cooled industrial induction motor," *IEEE Trans. Ind. Appl.*, vol. 39, no. 4, pp. 945–952, Jul./Aug. 2003.

[18] G. Kylander, "Temperature simulation of a 15kW induction machine operated at variable speed," in *ICEM92, Manchester*, Sep. 15–17, 1992.

[19] D. A. Staton, "Thermal computer aided design–Advancing the revolution in compact motors, in *IEEE IEMDC 2001*, Boston, MA, Jun. 2001.

[20] J. Mugglestone, S. J. Pickering, and D. Lampard, "Effect of geometry changes on the flow and heat transfer in the end region of a TEFC induction motor," in *9th IEE Intl. Conf. Electrical Machines & Drives*, Canterbury, UK, Sep. 1999.

[21] D. Staton, S. J. Pickering, and D. Lampard, "Recent advancement in the thermal design of electric motors," in *SMMA 2001 Fall Technical Con*, Durham, NC, Oct. 3–5, 2001.

[22] R. E. Steven, Electrical Machines and Power Electronics, Van Nostrand Reinhold, 1983, ISBN 0-442-30548-6

[23] M. S. Rajagopal, K. N. Seetharamu, and P. A. Aswathnarayana, "Transient thermal analysis of induction motors," *IEEE Trans. Energy Convers.*, vol. 13, no. 1, Mar. 1998.

[24] G. Champenois, D. Roye, and D. S. Zhu, "Electrical and thermal performance prediction in inverter—Fed squirrel cage induction motor drives," *Electr. Mach. Power Syst.*, vol. 22, no. 3, pp. 335–370, 1994.

[25] J. T. Boys and M. J. Miles, "Empirical thermal model for inverter-driven cage induction machines," *IEE Proc. Electr. Power Appl.*, vol. 141, no. 6, pp. 360–372, Nov. 1995.

[26] S. Mezani, N. Talorabet, and B. Laporte, "A combined electromagnetic and thermal analysis of induction motors," *IEEE Trans. Magn.*, vol. 41, no. 5, pp. 1572–1575, May 2005.

[27] D. G. Dorrrell, D. A. Staton, J. Hahout, D. Hawkins, and M. I. McGilp, "Linked electromagnetic and thermal modelling of a permanent magnet motor," in *PEMD 2006*, Dublin, Apr. 2006.

[28] J. R. Hendershot and T. J. E. Miller, Design of Brushless Permanent-Magnet Motors, Clarendon Press, 1994, ISBN 0-19-859389-9

[29] Y. See, S. Hahn, and S. Kauh, "Thermal analysis of induction motor with forced cooling channels," *IEEE Trans. Magn.*, vol. 36, no. 4, pp. 1398–1402, Jul. 2000.

[30] J. F. Trigeol, Y. Bertin, and P. Lagonotte, "Thermal modeling of an induction machine through the association of two numerical approaches," *IEEE Trans. Energy Convers.*, vol. 21, no. 2, pp. 314–323, Jun. 2006.

[31] D. Staton, A. Boglietti, and A. Cavagnino, "Solving the more difficult aspects of electric motor thermal analysis, in small and medium size industrial induction motors," *IEEE Trans. Energy Convers.*, vol. 20, no. 3, pp. 620–628, Sep. 2005.

[32] C. Mejuto, M. Mueller, M. Shanel, A. Mebarki, and D. Staton, "Thermal modelling investigation of heat paths due to iron losses in synchronous machines," in *4th Int. Conf. on Power Electronics, Machines & Drives*, York St John University College, York, UK, Apr. 4, 2008.

[33] A. Boglietti, A. Cavagnino, and D. Staton, "Determination of critical parameters in electrical machine thermal models," in *2007 IEEE Industry Applications Annual Meeting*, Sep. 23–27, 2007, pp. 73–80.

[34] S. N. Rea and S. E. West, "Thermal radiation from finned heat sinks," *IEEE Trans. Hybrids Packag.*, vol. 12, no. 2, pp. 115–117, Jun. 1976.

[35] M. F. Modest, *Radiative Heat Transfer*, Academic Press, 2003, ISBN 0125031637

[36] A. Cavagnino and D. Staton, "Convection heat transfer and flow calculations suitable for analytical modelling of electric machines," in *CD Conf. Rec. IECON06*, Paris, France, Nov. 6–10, 2006, pp. 4841–4846.

[37] R. W. Fox, A. T. McDonald, and P. J. Pritchard, Introduction to Fluid Mechanics. John Wiley & Sons, 2004, ISBN 0-471-37653-1

[38] I. E. Idlechik, *Handbook of Hydraulic Resistance—Coefficients of Local Resistance and of Friction.* NTIS, 1960.

[39] W. C. Osborne and C. G. Turner, Eds., *Woods Practical Guide to Fan Engineering,* 2nd ed. Woods of Colchester Ltd, Jun. 1960.

[40] D. A. Lightband and D. A. Bicknell, *The Direct Current Traction Motor: Its Design and Characteristics.* London: Business Books, 1970.

[41] J. L. Taylor, Calculating air flow through electrical machines, *Electrical Times,* Jul. 21, 1960.

[42] Cummins Generator Technologies, www.cumminsgeneratortechnologies.com

[43] W. L. Miranker, *Numerical Methods for Stiff Equations and Singular Perturbation Problems.* D. Reidel Publishing Company, 1979.

[44] D. J. Powell, "Modelling of high power density electrical machines for aerospace," Ph.D. thesis, University of Sheffield, UK, May 2003.

[45] M. Shanel, S. J. Pickering, and D. Lampard, "Application of computational fluid dynamics to the cooling of salient pole electrical machines," in *Proc. of ICEM 2000, Int. Conf. Electr. Mach,* Espoo, Finland, Aug. 2000, vol. 1, pp. 338–342.

[46] S. J. Pickering, D. Lampard, and M. Shanel, "Modelling ventilation and cooling of the rotors of salient pole machines," in *IEMDC2001,* Cambridge, MA, Jun. 2001.

[47] M. Shanel, S. J. Pickering, and D. Lampard, "Conjugate heat transfer analysis of a salient pole rotor in an air cooled synchronous generator," in *IEMDC2003,* Madison, WI, Jun. 1–4, 2003.

[48] A. Boglietti and A. Cavagnino, "Analysis of endwinding cooling effects in TEFC induction motors," *IEEE Trans. Ind. Appl.,* vol. 43, no. 5, pp. 1214–1222, Sep/Oct. 2007.

[49] A. Boglietti, A. Cavagnino, and D. Staton, "TEFC induction thermal models: A parameters sensitivity analysis," *IEEE Trans. Ind. Appl.,* vol. 41, no. 3, pp. 756–763, May/Jun. 2005.

[50] M. A. Valenzuela and J. A. Tapia, "Heat transfer and thermal design of finned frames for TEFC variable speed motors," in *CD Conf. Rec. IECON06,* Paris, 2006, pp. 6–10.

[51] D. Staton, "Thermal analysis of electric motors and generators," Tutorial course, in *IEEE IAS Annual Meeting,* Chicago, 2001.

[52] I. J. Perez and J. K. Kassakian, "A stationary thermal model for smooth air-gap rotating machines," *Electr. Mach. Electrom.,* vol. 3, nos. 3–4, pp. 285–303, 1979.

[53] IEC 60085:2004, "Electrical insulation—Thermal evaluation and designation," 3rd ed., International Electrotechnical Commission.

[54] NEMA standard MG-1: Motors and Generators, 2014.

[55] M. A. Laughton and D. F. Warne, Eds., *Electrical Engineer's Reference Book,* 16th ed. Newnes, 2004.

[56] D. G. Fink and W. H. Beaty, Eds., *Standard Handbook for Electrical Engineers,* 11th ed. McGraw Hill, 1978.

[57] P. Mellor, R. Wrobel, and N. Simpson "AC losses in high frequency electrical machine windings formed from large section conductors," in *IEEE ECCE 2014,* Pittsburgh, PA, Sep. 2014.

[58] A. Boglietti, A. Cavagnino, D. Staton, M. Shanel, M. Mueller, and C. Mejuto, "Evolution and modern approaches for thermal analysis of electrical machines," *IEEE Trans. Ind. Electron.,* vol. 56, no. 3, pp. 871–882, 2009.

[59] http://www.elantas.com

[60] http://www.dupont.com/products-and-services/

[61] http://www.victrex.com/PEEK

[62] J. S. Hsu, C. W. Ayers, C. L. Coomer, R. H. Wiles, S. L. Campbell, K. T. Lowe, and R. T. Michelhaugh, "Report on Toyota/Prius motor torque capability, torque property, no-load back emf, and mechanical losses," Oak Ridge Institute for Science and Education, Oak Ridge National Laboratory Report #ORNL/TM-2004/185.

[63] T. A. Burress, S. L. Campbell, C. Coomer, C. W. Ayers, A. A. Wereszczak, J. P. Cunningham, L. D. Marlino, L. E. Seiber, H.-T. Lin, "Evaluation of the 2010 Toyota Prius hybrid synergy drive system," Oak Ridge Institute for Science and Education, Oak Ridge National Laboratory Report #ORNL/TM-2010/253

[64] http://www.techmasterinc.com/engineerweb/yaskawamotor/
[65] http://www.servo-drive.cz/inverted_roller_screw_servoactuator.php
[66] http://www.proteanelectric.com
[67] D. Staton, "Multiphysics analysis of electric machines for traction applications considering complex duty cycles," Special session, in *IEEE ECCE 2014*, Pittsburgh, 2014.
[68] A. Boglietti, A. Cavagnino, M. Pervis, and A. Vallan, "Evaluation of radiation thermal resistances in industrial motors," *IEEE Trans. Ind. Appl.*, vol. 42, no. 3, pp. 688–693, 2006.
[69] http://www.equipmake.co.uk
[70] T. Burress, "Benchmarking state-of-the-art technologies," Oak Ridge National Laboratory, May 2013. [Online]. http://energy.gov/sites/prod/files/2014/03/f13/ape006_burress_2013_o.pdf
[71] T. Woolmer, "Wheel-hub motor cooling," U.S. Patent Appl. 20130187492 A1.
[72] S. H. Swales, P. F. Turnbull, B. Schulze, F. R. Poskie, W. J. Omell, W. C. Deneszczuk, "Oil cooled motor/generator for an automotive powertrain," U.S. Patent Appl. US 8169110 B2.
[73] L. Fedoseyev and E. M. Pearce, Jr., "Rotor assembly with heat pipe cooling system," U.S. Patent Appl. 2014 /0368064 A1.
[74] G. C. Stone, I. Culbert, E. A. Boulter, and H. Dhirani, Electrical insulation for rotating machines: Design, evaluation, aging, testing, and repair, in *IEEE Press Series on Power Engineering*, 2nd ed. John Wiley & Sons, 2014.
[75] E. Driessens, "The submerged double jet impingement (SDJI) method for thermal testing of packages," *Electronics Cooling*, vol. 7, no. 2, May 2001.
[76] htttp://www.bayer.com/
[77] http://multimedia.3m.com/mws/media/648880/fluorinert-electronic-liquid-fc-40.pdf?fn=prodinfo_FC40.pdf

CHAPTER 5

AUTOMATED OPTIMIZATION FOR ELECTRIC MACHINES

5.1 INTRODUCTION

The optimized design of electrical machines is a nonlinear multi-objective problem. A substantial amount of literature has been devoted to this subject over the last decade [1]. In electrical machine design, objectives such as highest efficiency, lowest cost, and minimum weight of active materials have to be simultaneously met. Further, this electromagnetic problem should be solved with consideration of the mechanical, thermal, and materials constraints. The results of such a multi-objective optimization problem are interpreted with the use of Pareto fronts (Figure 5.1). A Pareto front is a geometrical entity along which improvement in objective can only be achieved through a deterioration in another objective. The Pareto front can aid in identifying a family of best designs. It also provides a fair basis for the comparison of different machine topologies. Comparison at only one design point might lead to erroneous conclusions (Figure 5.2).

Electric machine designers today rely on rules of thumb for initial sizing and topology selection. For example, it is believed that the 12/10 slot pole configuration is a good choice for brushless permanent magnet (PM) machines. This is true provided that core loss are not significant, eccentricity can be maintained low, and effect of high radial forces can be dampened in the system. It is also widely held that a synchronous reluctance motor can have higher specific torque and efficiency than its induction counterpart. This is true, provided that, the saliency ratio of the former is very high (between 7 and 10), and stray load losses are kept low through material selection and process control. These "provided that" conditions may be hard to achieve in actual practice. Systematic design optimization enables taking into account many of these.

Multiphysics Simulation by Design for Electrical Machines, Power Electronics, and Drives, First Edition.
Marius Rosu, Ping Zhou, Dingsheng Lin, Dan Ionel, Mircea Popescu, Frede Blaabjerg, Vandana Rallabandi, and David Staton.

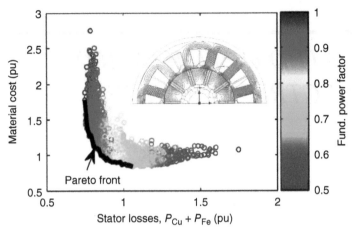

Figure 5.1   Pareto front in a multi-objective optimization problem. Finite element analysis is used for the performance calculation [2].

5.2   FORMULATING AN OPTIMIZATION PROBLEM

The steps in the process of formulating an optimization problem are as follows: identification of objective functions, selection of the input design variables, and definition of the constraints.

5.2.1   Objective Functions

Optimization problems can be single or multi-objective.

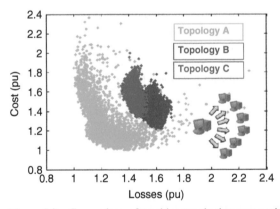

Figure 5.2   Comparison of machine topologies over a wide range of designs leading to the conclusion that Topology A has the minimum cost per loss among those compared. Parallel computing power is used to solve this large-scale optimization problem.

Single Objective Optimization

A single weighted objective function is formulated for N performance indices as follows:

$$f_0 = \sum_{n=1}^{N} w_n f_n(x) \tag{5.1}$$

where $w_1, w_2, \ldots, w_n$ are the weights.

The advantages of such a method are simple implementation and interpretation of results as compared with multi-objective optimization, however, correct choice of the weights is a challenge.

As an example, consider a generic optimization study for a permanent magnet motor. The objective function is selected to be the cost per point of efficiency as is defined as [3]

$$\frac{C_m}{\eta} = \frac{C_{Fe} m_{Fe} + C_{Cu} m_{Cu} + C_{PM} m_{PM}}{\eta} \tag{5.2}$$

where m_{Fe}, m_{Cu}, m_{PM} are the losses in the iron, copper, and permanent magnets, respectively, and C_{Fe}, C_{Cu}, C_{PM} are the weights.

Multi-objective Optimization

Rather than using a single objective function combining all performance indices, individual objective functions corresponding to each performance index are defined. The advantages are that no weights are needed, and a family of best compromise designs can be identified from a Pareto plot. The limitations of this approach are that the implementation is more complex, and the results might be hard to interpret. Examples of multiple objective functions that could be used in electric machine design are torque per unit mass, torque ripple, etc.

$$f_T = \frac{T_{em}}{W_{PM}}, \quad f_{ripple} = \frac{\max(T_{em}) - \min(T_{em})}{T_{em}}, \text{ etc.} \tag{5.3}$$

5.2.2 Input Design Variables

In case of an electric machine, input design variables (independent variable) are generally geometric dimensions. An example is shown in Figure 5.3.

5.2.3 Steps in Systematic Design

1. Establishing the specification
2. Selection of motor topology
3. Initial sizing based on analytical calculations
4. Optimal design including electromagnetic, mechanical, and thermal constraints including estimation of performance and cost
5. Studying sensitivity to manufacturing

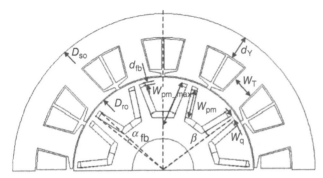

Figure 5.3 Independent variables in an example interior PM machine [4].

6. Noise and vibration analysis
7. Comparing topologies. This might lead to findings, which necessitate returning to step 2. The process is iterative.
8. Making a decision.

5.3 OPTIMIZATION METHODS

Optimal search algorithms can be broadly divided into deterministic and stochastic methods. The former finds the optimum solution algorithmically, while the stochastic methods conduct a random exploration of the solution space. Furthermore, optimization methods are also classed depending upon whether they have scalar or vector objective functions. Multi-objective optimization problems are typically formulated with vector objective functions, while the scalar formulation is generally used for single objective optimization.

For scalar objective functions, examples of deterministic methods include sequential unconstrained minimization technique (SUMT), which converts a constrained optimization problem into one which is unconstrained by the use of penalty functions [5]. Following this, nonlinear programming is employed for optimization. Some applications of this method in electric machine design are discussed in [6–11]. Other deterministic methods are Hooke–Jeeves method, interval branch and bound method, etc.

Stochastic methods such as Genetic Algorithm (GA) [12], Simulated Annealing [13], Particle Swarm Optimization (PSO) [14], and Differential Evolution (DE). require more candidate designs, but generally manage to find the global optima, as opposed to deterministic methods which might be trapped by local optima. Genetic Algorithm has been reported for the design optimization of a synchronous motor [15], and for a coreless axial flux permanent magnet machine [16], while PSO has been used for the design optimization of permanent magnet synchronous machine [17] and for torque ripple minimization of an interior PM machine [18].

For optimization algorithms pertaining to vector objective functions, common algorithms include multi-objective GA (MOGA), multi-objective PSO (MOPSO), multi-objective DE (MODE), and methods based on Design of Experiments (DOE). A MOGA algorithm was used for flux barrier optimization of synchronous reluctance machines in [19]. MOGA, MODE, and Multi-objective Simulated Annealing were employed for the design of a multi-layer synchronous reluctance machine, and compared. It was found that MODE had the best convergence time and repeatability of results [20].

A multi-objective optimization evolutionary algorithm based on decomposition (MOEA/D) was proposed [21] and applied for the design of an interior PM machine in [22]. This method decomposes the multi-objective optimization problem into several scalar problems, and information from adjacent problems is utilized to achieve faster convergence. An algorithm combing DE with multi-objective optimization was applied for the optimization of IPM motors by the authors [23].

Stochastic methods require a large number of candidate models, necessitating the use of surrogate modeling which approximates the relationship between input and output. One such technique called computationally efficient finite element analysis (CEFEA) was proposed by the authors in [24, 25]. Other approaches include Kriging-based methodologies, which reduce the number of required evaluations of the objective function. These methods are applicable to smaller scale problems due to the prohibitively high computational resources required for inversion of the correlation matrix for larger problems. A points aggregation-based approach to mitigate this limitation is proposed in [26].

Figure 5.4 illustrates the two main approaches for systematic design optimization: DOE and differential evolution, discussed in this chapter in some detail.

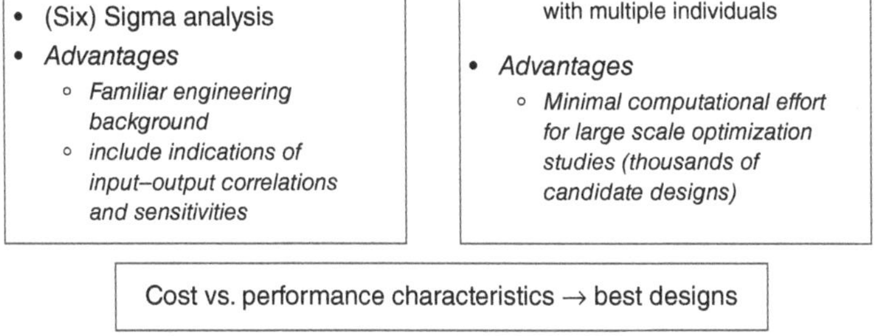

Machine specification and initial design

• Design of experiments (DOE)	• Computational intelligence
○ Parameter correlation	○ Differential evolution (DE); others: GA, PSO, etc.
○ Response surface (RS)	
○ Estimated performance based on RS	○ Inspired from natural selection
• (Six) Sigma analysis	○ Generations / populations, each with multiple individuals
• *Advantages*	• *Advantages*
○ *Familiar engineering background*	○ *Minimal computational effort for large scale optimization studies (thousands of candidate designs)*
○ *include indications of input–output correlations and sensitivities*	

Cost vs. performance characteristics → best designs

Figure 5.4 Two major approaches for automated design optimization. Combinations between the two are possible.

5.4 DESIGN OF EXPERIMENTS AND RESPONSE SURFACE METHODS

5.4.1 Overview

The DOE and Response Surface (RS) methods are techniques, which are employed for the establishment of a relationship between the inputs and outputs of a system. This relationship could be used for optimization. In the case of an electric machine, the designer is interested in the performances, say, the reduction of torque ripple and cost and the improvement of efficiency. The performances are defined as responses, which are functions of independent geometric variables (e.g., air gap, stator tooth width, and rotor pole arc). The independent geometric variables (input design variables) constitute the inputs. The relationship between the responses (y) and input design variables (x) can be established by a series of experiments. DOE is the process of planning experiments in a systematic way so that data collected can be used to fit appropriate models. Prior to conducting the experiments, levels are defined for the input design variables. Full factorial designs refer to a series of experiments wherein the values of the input design variables are fixed at the predetermined levels. For instance, in a two-level, two-factor full factorial design, with independent input variables a and b, the runs are as in Table 5.1. The data from the experiments can be analyzed with the help of main and interaction effects. The mathematical developments from this section follow the approach described in [27–29]. The main effects are calculated as

$$A = \frac{y_a^+ - y_a^-}{n} \tag{5.4}$$

where y_a^+ and y_a^- are the sum of the values of the outputs with the factor a at high and low levels (as shown in Table 5.1), respectively and n, the number of runs. The main effect factors for all input variables are defined similarly.

The interaction effect AB is defined as the difference between the effects of a when b is held at a high level and a low level.

$$AB = \frac{e_b^+ - e_b^-}{n} \tag{5.5}$$

TABLE 5.1 Runs in a two-level full factorial design. + and − refer to the minimum and maximum values of the input design variables, respectively

Run	a	b
1	−	−
2	−	+
3	+	−
4	+	+

TABLE 5.2   Runs in a two-level, three-factor full factorial design

Run	a	b	c	ab	bc	ac	abc
1	−	−	−	+	+	+	−
2	+	−	−	−	+	−	+
3	−	+	−	−	−	+	+
4	+	+	−	+	−	−	−
5	−	−	+	+	−	−	+
6	+	−	+	−	−	+	−
7	−	+	+	−	+	−	−
8	+	+	+	+	+	+	+

where e_b^+ is the sum of effects of a with b held high and e_b^- the sum of the effects of a with b held low.

The effect of a is defined as the difference between the outputs when a is high and when a is low, at a fixed value of b.

5.4.2   Fractional Factorial Methods

For systems where a large number of input variables are expected to affect the output, the number of full factorial runs becomes prohibitively high. This is when fractional factorial method can be used. Consider Table 5.2 which shows the runs required for a three-factor, two-level DOE. A total of eight runs are needed to estimate all the main and interaction effects.

The main effect A is

$$A = \tfrac{1}{8}(-y_1 + y_2 - y_3 + y_4 - y_5 + y_6 - y_7 + y_8) \tag{5.6}$$

The interaction effect AB is

$$AB = \tfrac{1}{8}(y_1 - y_2 - y_3 + y_4 + y_5 - y_6 - y_7 + y_8) \tag{5.7}$$

The interaction effect ABC is

$$ABC = \tfrac{1}{8}(-y_1 + y_2 + y_3 - y_4 + y_5 - y_6 - y_7 + y_8) \tag{5.8}$$

where $y_1, y_2, \ldots, y_8$ represent the outputs of runs 1–8.

A point to be noted is that the signs of AB, ABC, etc., which are the signs of the coefficients in equations (5.6–5.8) are also obtained by multiplying the signs of the appropriate a, b, and c columns. In case one wants to do a half-factorial run for this experiment, it would mean deletion of four rows. As an example, consider that only the second, third, fifth, and eighth runs of Table 5.2 are retained. In this case, the main and interaction effects are re-written as

$$A = \tfrac{1}{4}(y_2 - y_3 - y_5 + y_8)$$

$$B = \tfrac{1}{4}(-y_2 + y_3 - y_5 + y_8)$$

$$C = \tfrac{1}{4}(-y_2 - y_3 + y_5 + y_8)$$

$$AB = \tfrac{1}{4}(-y_2 - y_3 + y_5 + y_8) \tag{5.9}$$

$$BC = \tfrac{1}{4}(y_2 - y_3 - y_5 + y_8)$$

$$AC = \tfrac{1}{4}(-y_2 + y_3 - y_5 + y_8)$$

$$ABC = \tfrac{1}{4}(-y_2 + y_3 + y_5 + y_8)$$

Thus, $A = BC$, $B = AC$, and $C = AB$, which means it is not possible to distinguish between these effects. In other words, the main effects are confounded with second-order interaction effects in a fractional factorial run. Domain knowledge would help one identify which of the input variables is likely to have a significant effect on the output, as well as the highest order of interactions that need to be considered. Accordingly, the decision whether to run a fractional factorial or not could be taken.

5.4.3 Response Surface Model

The relation between inputs and outputs of a system is given as

$$y_n = f(x_1, x_2, \ldots, x_{Dv}) + \in \tag{5.10}$$

The input design variables are called natural variables if expressed in the natural unit for measurements (length in mm and angle in degrees). The natural variables may be transformed to coded variables, dimensionless by definition, with a mean of zero and the same standard deviation as follows:

$$C = \begin{bmatrix} c_1, c_2, \ldots, c_{Dv} \end{bmatrix} = \frac{X - (X_{\min} + X_{\max})/2}{(X_{\max} - X_{\min})/2} \tag{5.11}$$

The function may be reformulated in terms of the coded variables using a first- or second-order least square model given as

$$y = \beta_0 + \sum_{i=1}^{Dv} \beta_i c_i + \sum_{i=1}^{Dv} \beta_{ii} c_{ii}^2 + \sum_{i=1}^{Dv} \sum_{j=i+1}^{Dv} \beta_{ij} c_i c_j \tag{5.12}$$

where β_0, β_i, β_{ii}, β_{ij} are the regression coefficients. The fitted model is called a response surface model (RSM). Second-order RSMs are widely used. A second-order

RSM for two design variables is given as

$$y = \beta_0 + \beta_1 c_1 + \beta_{11} c_1^2 + \beta_{12} c_1 c_2 + \beta_2 c_2 + \beta_{22} c_2^2 + \in \qquad (5.13)$$

Evaluation of the main and interaction effects help determine the type of RSM that may be fit. If only the main effects are important, then the interaction terms β_{12} may be eliminated. From prior knowledge, if the system is expected to be nonlinear, more than two levels are needed. Even in such systems, a two-level DOE may be run in the initial screening phase to determine which input variables would have an effect on the output.

The minimum number of experiments (N) required to determine first- and second-order coefficients of an RSM is given as

$$N = 1 + D_v + \frac{D_v(1 + D_v)}{2} \qquad (5.14)$$

Further, in order to fit a second-order response surface, at least three levels of the input design parameters must be considered. A procedure for optimization using DOE and RSM is shown in Figure 5.5, and an instance of a response plot is shown in Figure 5.6, while Figure 5.7 shows an instance of sensitivity analysis.

Some commonly used DOE approaches include the central composite design which considers five levels of the coded input design variables $[-\alpha, -1, 0, 1, \alpha]$, where the value of α depends on the type of method (Central Composite Circumscribed, Central Composite Inscribed, Central Composite Face Centered).

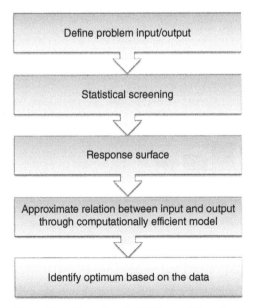

Figure 5.5   Algorithm for DOE and RS methods.

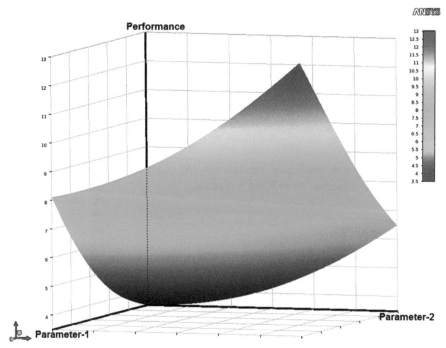

Figure 5.6 Example response surface plot.

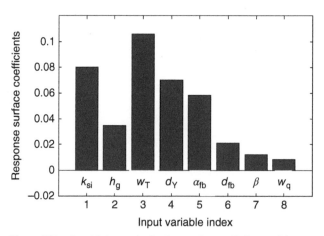

Figure 5.7 Sensitivity analysis: first-order coefficients of the response surface, which can be used to identify which independent input variables affect the output. This applies to the machine topology illustrated in Figure 5.3 [28].

5.5   DIFFERENTIAL EVOLUTION

Differential evolution (DE), a method for multi-objective optimization was first proposed by Storn and Price in 1995 [30]. A DE algorithm contains a number of g_{max} generations, each with a population of N_p individual designs. The number of generations is taken as 10 times the number of independent variables. The algorithm consists of the following steps: initialization, mutation, cross-over, and selection. Before the process of initialization, the bounds of the design space must be defined (X_L and X_U), the suffixes indicating lower and upper bounds, respectively. The mathematical development follows from [28, 31]

5.5.1   Initialization

A random generator is used to obtain the initial values of the jth design variable of the ith design in the first generation as follows:

$$x_{j,i,1} = rand_j\,(0,1) \cdot (x_{jU} - x_{jL}) + x_{jL} \tag{5.15}$$

Following the creation of a generation containing N_p designs (the vector containing all D design parameters is designated by $X_{i,\,g}$, where i is the population index and g the generation index and is called the target vector), the design objectives and constraints are evaluated.

5.5.2   Mutation

Each design variable undergoes a mutation, which expands the search space. For the jth design variable, the mutation is carried out as follows:

$$v_{j,i,g} = x_{j,r1,g} + F(x_{j,r2,g} - x_{j,r3,g}) \tag{5.16}$$

The indices $r1$, $r2$, and $r3$ are all distinct, and not equal to i. The scale factor, F is between $(0,1+)$, that is, a positive value with no upper limit. A critical value of F was proposed [32] to be

$$F_{crit} = \sqrt{\frac{\left(1 - \frac{CR}{2}\right)}{N_p}} \tag{5.17}$$

where CR is the cross-over probability (explained subsequently). Larger values of F lead to better population diversity and algorithm convergence.

5.5.3   Cross-Over

This process builds trial designs $U_{i,g}$ out of designs copied from $X_{i,g}$ and $V_{i,g}$.

$$U_{i,\,g} = \begin{cases} V_{i,g} & \text{if } RAND\,(0,1) \leq CR \\ X_{i,g} & \text{otherwise} \end{cases} \tag{5.18}$$

where CR is the cross-over probability (the use of the uppercase *RAND* indicates that a random value is generated for each of the design variables).

5.5.4 Selection

In this step, the objective function evaluated for the trial designs is compared against that evaluated for the target vector, to generate a better target vector for the next generation, that is,

$$X_{i,\,g+1} = \begin{cases} U_{i,g} \text{ if } f(U_{i,g}) \le f(X_{i,g}) \\ X_{i,g} \text{ otherwise} \end{cases} \tag{5.19}$$

The processes of mutation, cross-over, and selection are repeated till the satisfaction of the stopping criteria, typically based on setting a maximum number of generations. Figure 5.8 illustrates the algorithm for DE, and Figure 5.9 the gradual population of the Pareto space.

5.6 FIRST EXAMPLE: OPTIMIZATION OF AN ULTRA HIGH TORQUE DENSITY PM MOTOR FOR FORMULA E RACING CARS: SELECTION OF BEST COMPROMISE DESIGNS

5.6.1 Problem Formulation

The optimization of a record-braking torque dense PM spoke type synchronous motor with an 18-slot, 16-pole topology for Formula E racing cars was reported in [34] and is summarized in this section.

There are eight geometric **independent design variables** (Table 5.3 and Figure 5.10). **Two concurrent objectives** were employed for minimizing (a) machine weight and (b) losses for a rated torque of 110 Nm at 6000 rpm, that is, approximately 100 hp.

The constraints are: (1) Torque ripple is maintained below 5% and (2) flux density in the PMs is maintained above 20% of Br.

5.6.2 Optimization Methodology

Combined multi-objective optimization with DE (vector objective function) is employed. Following DE, sensitivity analysis using Pearson correlation coefficients is performed for 100 designs on the Pareto front, to better understand the relationship between the inputs and objective functions. This is another example of going back and forth between DE and DOE type methods. The Pearson correlation $[cor(x, y)]$ between inputs x and objectives y may be calculated as

$$cor(x, y) = \frac{cov(x, y)}{\sigma_x \sigma_y}$$

$$cov(x, y) = \frac{\sum_{i=1}^{N} (x_i - x_m)(y_i - y_m)}{N} \tag{5.20}$$

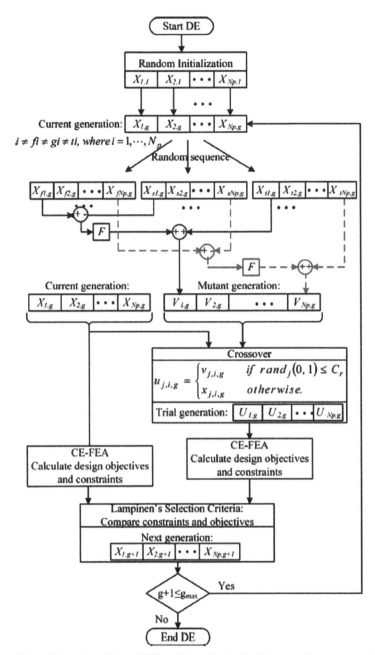

Figure 5.8   Algorithm of differential evolution. In this case, electromagnetic CEFEA is used for fast evaluations of the objective functions [33].

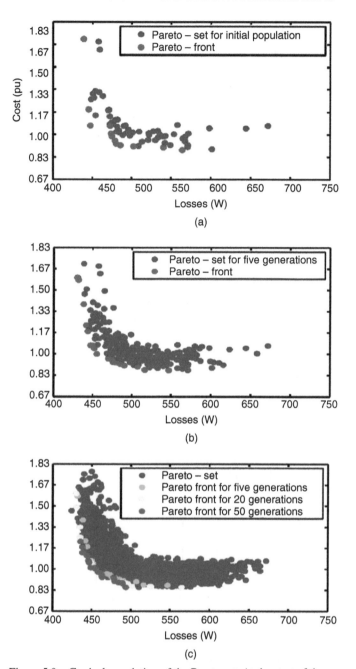

Figure 5.9   Gradual population of the Pareto set. At the start of the process, the Pareto set is sparse, and gets gradually more populated as the solution progresses [31].

TABLE 5.3   Independent design variables

Variable	Minimum	Maximum
Ratio of inner to outer radius	0.6	0.75
Air-gap length (h_g) (mm)	0.7	2.5
Ratio of stator tooth width to slot pitch	0.45	0.75
Ratio of PM height to the maximum PM height	0.55	0.95
Ratio of PM width to pole pitch	0.2	0.6
Ratio of rotor slot opening (w_{br}) to PM width	0.35	0.65
Depth of bridge (d_{br}) (mm)	1.5	3
Stator back iron width (h_y) (mm)	7	15

where $cov(x, y)$ is the covariance; x_m and y_m are the mean values of input variables and objectives, respectively; σ_x, σ_y, the standard deviations of x and y; and N the number of observations.

5.6.3   Results

The results of Figure 5.11 indicate that the machine mass is strongly affected by the choice of design parameters.

Rules of thumb, based on analytical calculations state that the split ratio (ratio of inner to outer radius) is in the range of 0.5–0.55, and the ratio of PM thickness to pole pitch is between 0.4 and 0.5 for high power density 18/16 pole machines [36]. However, these rules are based on studies, which neglect nonlinearity of the core,

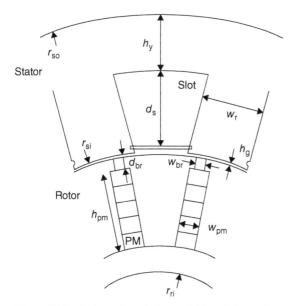

Figure 5.10   Independent design variables of the spoke type IPM machine [35].

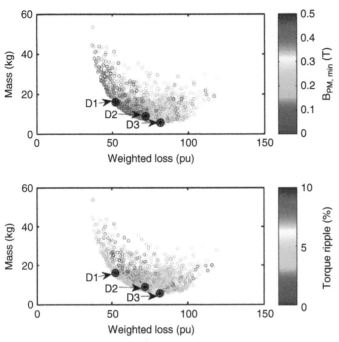

Figure 5.11   Results of the optimization study. D1, D2, and D3 represent best compromise designs selected out of 3400 candidates [35].

various loss components, and the complexity of the machine geometry. The values of these parameters arrived at as a consequence of this study (Figure 5.12) are different from those reported previously, emphasizing the need for nonlinear calculations.

5.7   SECOND EXAMPLE: SINGLE OBJECTIVE OPTIMIZATION OF A RANGE OF PERMANENT MAGNET SYNCHRONOUS MACHINE (PMSMS) RATED BETWEEN 1 kW AND 1 MW DERIVATION OF DESIGN PROPORTIONS AND RECOMMENDATIONS

5.7.1   Problem Formulation

This problem considers PM brushless motor designs with two stator slots per pole per phase and a single layer distributed winding over a very large range of power (Figure 5.13a). All the motor designs use typical grades of PMs (NdFeB with Br of 1.23 T) and lamination grade (M19), and have the same frequency of 120 Hz [3]. The study is summarized here.

A single objective function, cost per unit efficiency was employed in the study. **Two constraints**, that is, (a) peak-to-peak torque ripple of 20%, (b) minimum flux density in NdFeB PM > 0.4, **eight independent design variables** (Table 5.4) are considered.

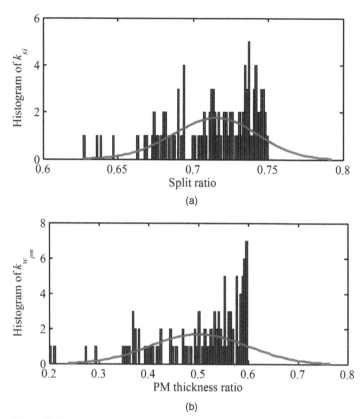

Figure 5.12   (a) The distribution of values of the split ratio, and (b) the ratio of PM thickness to pole pitch for 100 designs on the Pareto front [35].

DOE was used for sensitivity analysis, to establish the important independent variables affecting the output (Figure 5.13b)

5.7.2   Optimization Methodology

Differential evolution with 60 generations, each with a population of 70 individuals is employed for optimization. The stopping criteria used were: (a) stopping criteria in solution space—the objective function value does not improve beyond 5% across 20 generations and (b) the value of the average normalized distance of each design variable (i.e., measure of how different the individuals in a generation are from the best individuals) reduces below 10%.

5.7.3   Results

The study led to many interesting findings, for instance, it was seen that the cost per unit efficiency increases exponentially with the power rating. Also seen is that

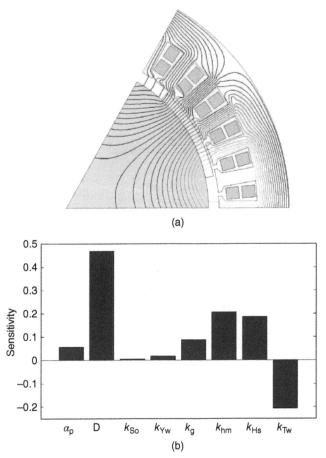

Figure 5.13 (a) Cross section of the 6-pole PMSM considered for study and (b) sensitivity of the objective function to the different independent input variables [3].

TABLE 5.4 Independent design variables considered in the study

Variable	Minimum	Maximum
Diameter (mm)	50	500
Ratio of air gap to minimum air gap	1	3
Ratio of stator tooth width to slot pitch	0.3	0.7
Ratio of slot opening to slot width	0.05	0.9
Ratio of stator slot height to pole pitch	0.02	1.5
Ratio of stator yoke width to half pole pitch	0.4	0.8
Ratio of PM height to air-gap length	1.5	10
PM pole arc to pole pitch ratio	0.6	1

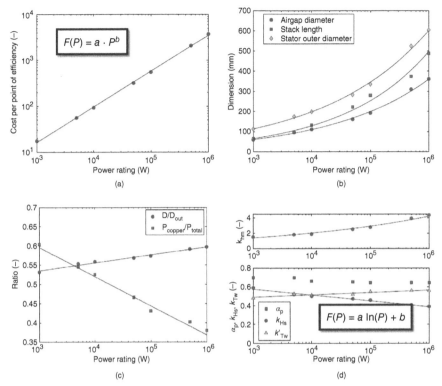

Figure 5.14 Results of the optimization study. Optimum values of design variables such as rotor pole arc, height of stator yoke, and width of stator tooth vary nonlinearly with the power rating [3].

the optimum split ratio increases with the increase in power rating, and machines of higher rating have a lower ratio of copper loss to total loss. The independent variables can be fit as a function of power using the results of Figure 5.14 to optimize PMSM designs across a wide range of ratings.

5.8 THIRD EXAMPLE: TWO- AND THREE-OBJECTIVE FUNCTION OPTIMIZATION OF A SYNCHRONOUS RELUCTANCE (SYNREL) AND PM ASSISTED SYNCHRONOUS RELUCTANCE MOTOR

5.8.1 Problem Formulation

A **multi-objective** optimization of the rotor geometry of a 36-slot, 4-pole, 4-barrier, 10-hp synchronous reluctance machine is done [37, 38]. **Two different problems** are formulated: the first being, a two-objective function problem, with a single constraint.

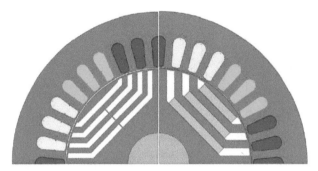

Figure 5.15   Cross section of the studied machine without and with PMs in the barriers. Nine independent variables which relate to the shape of the rotor flux barriers are considered [38].

In this problem, the objectives for minimization are badness (defined as the ratio of the square root of total loss to the torque) and torque ripple. **The constraint** considered is that power factor is above 0.7.

The second case considers a three-objective unconstrained optimization. The objectives include torque ripple, badness, and power factor. Power factor is multiplied by −1 to convert the problem into a minimization problem.

Nine geometrical ratios, relating to the shape of the rotor barriers were considered as independent variables for this study. Only the rotor geometry was optimized for a given stator. In an extension of this problem, permanent magnets were added to the rotor flux barriers. Both rotors, with and without permanent magnets, are illustrated in Figure 5.15.

5.8.2   Optimization Methodology

Multi-objective optimization combined with differential evolution is formulated and employed. The number of generations were 51, and 100 designs per generation were used.

Computationally efficient FEA which required a minimum number of magnetostatic solutions for performance estimation including torque ripple and core loss was used, in lieu of the time-consuming transient FEA (Figure 5.16).

5.8.3   Results

An interesting finding of the study was that of the two- and three-objective function problems, it was observed that the latter converged much faster, in the 22nd generation, which can also be inferred from the sequential ID of the optimum designs (Figure 5.17). Figure 5.18 shows the Pareto region identified by the three-objective function problem.

The studies also conclusively prove that though the automated optimal design procedure leads to a machine with high specific power, efficiency, and low torque ripple, the low power factor is an inherent limitation of this topology.

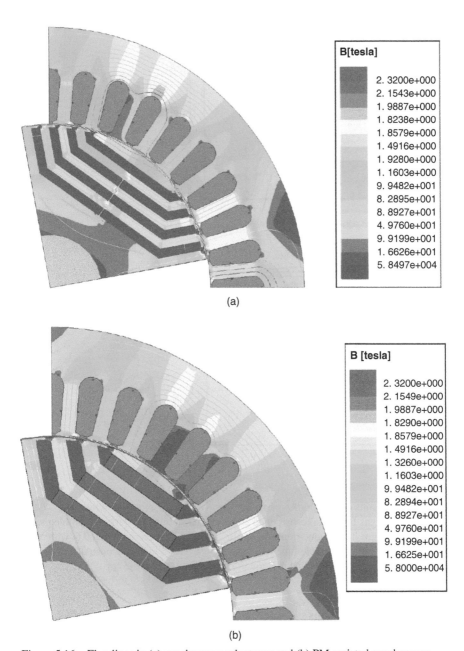

(a)

(b)

Figure 5.16   Flux lines in (a) synchronous reluctance and (b) PM assisted synchronous reluctance machine [37, 38].

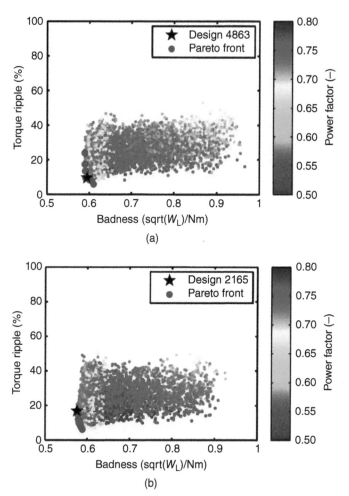

Figure 5.17 Results of the optimization study showing the performance of the best compromise designs for the (a) constrained and (b) unconstrained optimization problems. The optimal designs have torque ripple below 20%, and a power factor of 0.5–0.6, with badness of 0.6. The lower index of the best compromise design in the unconstrained optimization problem indicates that the solution was arrived at faster [37].

As an extension of this study, ferrite permanent magnets were added in the rotor barriers in order to investigate their effect. This topology is known as a PM assisted SynRel (PMaSynRel). The same optimization methods were used. The results of the study (Figure 5.19) indicate that for a 10-hp machine, the power factor can be improved from 0.75 to 0.95 by the addition of PMs to the rotor. In addition, the specific power is increased by over 30%, and the efficiency is improved too. No objective function which captured the cost of the PMs was used in the optimization, therefore the algorithm yielded machine designs which were relatively "PM heavy."

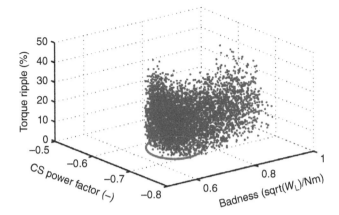

Figure 5.18   Pareto surface for the unconstrained problem [37].

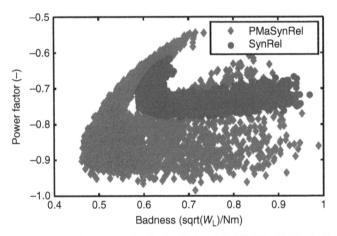

Figure 5.19   Pareto sets for the SynRel and the PMaSynRel including 10,000 candidate designs. The PMaSynRel designs can achieve lower badness and higher power factor [38].

5.9   FOURTH EXAMPLE: MULTI-OBJECTIVE OPTIMIZATION OF PM MACHINES COMBINING DOE AND DE METHODS

5.9.1   Problem Formulation

Permanent magnet machines with 12 slots, 8 and 10 poles were studied, including designs with both ferrite and NdFeB PMs [33]. The studied machines also had different PM arrangements, that is, V-type IPM, flat-bar IPM, and spoke type IPM (Figure 5.20). **Three objectives functions** were considered: minimum total material cost,

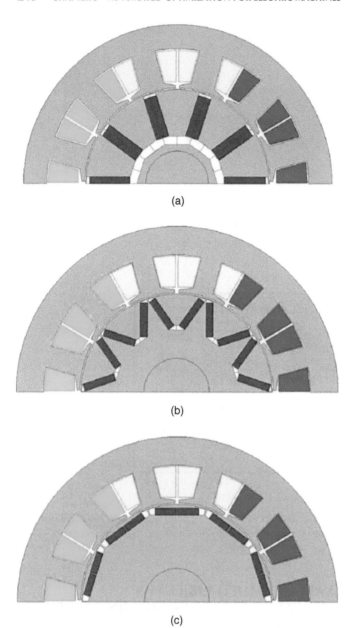

(a)

(b)

(c)

Figure 5.20   Cross sections of the studied IPM machines: (a) Spoke PM, (b) V-type, and (c) flat bar [33].

power losses, and torque ripple.

$$y_1 = \min(P_{Cu} + P_{Fe} + P_{PM} + P_{me})$$
$$y_2 = \min(5m_{pm} + 8m_{Cu} + 1m_{Fe}) \text{ for ferrite}$$
$$y_2 = \min(65m_{pm} + 8m_{Cu} + 1m_{Fe}) \text{ for NdFeB PMs}$$
$$\text{and}$$
$$y_3 = \min\left(\frac{T_{\max} - T_{\min}}{T_{\text{avg}}}\right)$$

(5.21)

In the objective function y_2, corresponding to material costs, different weighting factors were used to account for the price difference between NdFeB and ferrite magnets. Independent variables, numbering eight in total were considered for both the stator and rotor geometries.

5.9.2 Optimization Methodology

A DOE-based approach was initially used to identify the input variables with the most significant effect on the output. A central composite design, one of the most popular DOE methods, was employed to perform a sensitivity analysis, which revealed that only six of the eight input parameters had a sizeable effect on the objectives. This helped reduce the design space, leading to significant saving in time and computational resources. Following the DOE, a DE-based algorithm was used for the optimization.

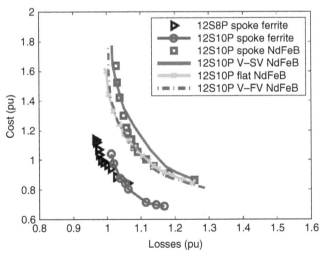

Figure 5.21   Pareto fronts for the studied machines. It is seen that all designs with NdFeB PMs, irrespective of rotor topology have similar performance, while the spoke-PM designs with ferrites have lower cost [33].

5.9.3 Results

The study revealed that for the rating considered (10 hp at 1800 rpm), similar cost and loss objectives can be met with any design of IPMs with NdFeB irrespective of the type of topology (V, flat bar, or spoke), as seen in Figure 5.21.

It is to be noted that the study did not consider manufacturing aspects. The study also indicated that comparable efficiency at lower cost (40%) can be achieved by the spoke-PM designs with ferrite PMs.

5.10 SUMMARY

The chapter discusses optimization as applied to electrical machine design. Some commonly used optimization methods are explained. Case studies illustrating the utility of systematic design optimization to compare different machine topologies, to develop design rules, and quantify the effect of different design features are included.

REFERENCES

[1] Y. Duan and D. M. Ionel, "A review of recent developments in electrical machine design optimization methods with a permanent-magnet synchronous motor benchmark study," *IEEE Trans. Ind. Appl.*, vol. 49, pp. 1268–1275, 2013.

[2] P. Zhang, D. M. Ionel, and N. A. O. Demerdash, "Saliency ratio and power factor of IPM motors with distributed windings optimally designed for high efficiency and low-cost applications," *IEEE Trans. Ind. Appl.*, vol. 52, pp. 4730–4739, 2016.

[3] Y. Duan and D. M. Ionel, "Non-linear scaling rules for brushless PM synchronous machines based on optimal design studies for a wide range of power ratings," in *IEEE ECCE*, 2012, pp. 2334–2341.

[4] P. Zhang, G. Y. Sizov, D. M. Ionel, and N. A. O. Demerdash, "Establishing the relative merits of interior and spoke-type permanent-magnet machines with ferrite or NdFeB through systematic design optimization," *IEEE Trans. Ind. Appl.*, vol. 51, pp. 2940–2948, 2015.

[5] N. Boules, "Design optimization of permanent magnet DC motors," *IEEE Trans. Ind. Appl.*, vol. 26, pp. 786–792, 1990.

[6] J. Legranger, G. Friedrich, S. Vivier, and J. C. Mipo, "Combination of finite-element and analytical models in the optimal multidomain design of machines: Application to an interior permanent-magnet starter generator," *IEEE Trans. Ind. Appl.*, vol. 46, pp. 232–239, 2010.

[7] K. Sun Bum, S. J. Park, and L. Jung Ho, "Optimum design criteria based on the rated watt of a synchronous reluctance motor using a coupled FEM and SUMT," *IEEE Trans. Magn.*, vol. 41, pp. 3970–3972, 2005.

[8] S. J. Mun, Y. H. Cho, and J. H. Lee, "Optimum design of synchronous reluctance motors based on torque/volume using finite-element method and sequential unconstrained minimization technique," *IEEE Trans. Magn.*, vol. 44, pp. 4143–4146, 2008.

[9] D. A. Kocabas, "Novel winding and core design for maximum reduction of harmonic magnetomotive force in AC motors," *IEEE Trans. Magn.*, vol. 45, pp. 735–746, 2009.

[10] L. Chedot, G. Friedrich, J. M. Biedinger, and P. Macret, "Integrated starter generator: The need for an optimal design and control approach. Application to a permanent magnet machine," *IEEE Trans. Ind. Appl.*, vol. 43, pp. 551–559, 2007.

[11] H. I. Lee and M. D. Noh, "Optimal design of radial-flux toroidally wound brushless DC machines," *IEEE Trans. Ind. Electron.*, vol. 58, pp. 444–449, 2011.

[12] S. N. Sivanandam and S. N. Deepa, *Introduction to Genetic Algorithms*. Springer Publishing Company, Incorporated, 2007.

[13] D. Bertsimas and J. Tsitsiklis, "Simulated annealing," *Statist. Sci.*, vol. 8, no. 1, pp. 10–15, 1993.

[14] J. Robinson and Y. Rahmat-Samii, "Particle swarm optimization in electromagnetics," *IEEE Trans. Antennas Propag.*, vol. 52, pp. 397–407, 2004.

[15] M. Celebi, "Weight optimisation of a salient pole synchronous generator by a new genetic algorithm validated by finite element analysis," *IET Electr. Power Appl.*, vol. 3, pp. 324–333, 2009.

[16] P. Virtič, M. Vražić, and G. Papa, "Design of an axial flux permanent magnet synchronous machine using analytical method and evolutionary optimization," *IEEE Trans. Energy Convers.*, vol. 31, pp. 150–158, 2016.

[17] J. H. Lee, J. W. Kim, J. Y. Song, D. W. Kim, Y. J. Kim, and S. Y. Jung, "Distance-based intelligent particle swarm optimization for optimal design of permanent magnet synchronous machine," *IEEE Trans. Magn.*, vol. 53, pp. 1–4, 2017.

[18] E. Aydin, L. Yingjie, I. Aydin, M. T. Aydemir, and B. Sarlioglu, "Minimization of torque ripples of interior permanent magnet synchronous motors by particle swarm optimization technique," in *ITEC*, 2015, pp. 1–6.

[19] G. Pellegrino, F. Cupertino, and C. Gerada, "Automatic design of synchronous reluctance motors focusing on barrier shape optimization," *IEEE Trans. Ind. Appl.*, vol. 51, pp. 1465–1474, 2015.

[20] F. Cupertino, G. Pellegrino, and C. Gerada, "Design of synchronous reluctance motors with multi-objective optimization algorithms," *IEEE Trans. Ind. Appl.*, vol. 50, pp. 3617–3627, 2014.

[21] Q. Zhang and H. Li, "MOEA/D: A multiobjective evolutionary algorithm based on decomposition," *IEEE Trans. Evolut. Comput.*, vol. 11, pp. 712–731, 2007.

[22] R. C. P. Silva, M. Li, T. Rahman, and D. A. Lowther, "Surrogate-based MOEA/D for electric motor design with scarce function evaluations," *IEEE Trans. Magn.*, vol. 53, no. 6, pp. 1–4, 2017.

[23] A. Fatemi, D. M. Ionel, N. A. O. Demerdash, and T. W. Nehl, "Fast multi-objective CMODE-Type optimization of PM machines using multicore desktop computers," *IEEE Trans. Ind. Appl.*, vol. 52, no. 4, pp. 2941–2950, 2016.

[24] Y. Wang, D. M. Ionel, and D. Staton, "Ultrafast steady-state multiphysics model for PM and synchronous reluctance machines," *IEEE Trans. Ind. Appl.*, vol. 51, pp. 3639–3646, 2015.

[25] D. M. Ionel and M. Popescu, "Ultrafast finite-element analysis of brushless PM machines based on space-time transformations," *IEEE Trans. Ind. Appl.*, vol. 47, pp. 744–753, 2011.

[26] Y. Li, S. Xiao, M. Rotaru, and J. Sykulski, "A kriging based optimization approach for large datasets exploiting points aggregation techniques," *IEEE Trans. Magn.*, vol. 53, no. 6, pp. 1–4, 2017.

[27] D. C. Montgomery, *Design and Analysis of Experiments*. John Wiley & Sons, 2008.

[28] P. Zhang, "A novel design optimization of a fault-tolerant AC permanent magnet machine-drive system," Ph.D. thesis, Marquette University, 2013.

[29] (2/27). *Fractional Factorial Designs—Duke Statistics*.

[30] R. Storn and K. Price, "Differential evolution—A simple and efficient heuristic for global optimization over continuous spaces," *J. Global Optim.*, vol. 11, pp. 341–359, 1997.

[31] Y. Wang, "Coupled electromagnetic and thermal analysis and design optimization of synchronous electric machines," MS, The University of Wisconsin-Milwaukee, 2014.

[32] K. Price, R. M. Storn, and J. A. Lampinen, *Differential Evolution: A Practical Approach to Global Optimization*, Springer, 2005.

[33] P. Zhang, G. Y. Sizov, D. M. Ionel, and N. A. O. Demerdash, "Establishing the relative merits of interior and spoke-type permanent-magnet machines with ferrite or NdFeB through systematic design optimization," *IEEE Trans. Ind. Appl.*, vol. 51, pp. 2940–2948, 2015.

[34] A. Fatemi, "Design optimization of permanent magnet machines over a target operating cycle using computationally efficient techniques," Ph.D. thesis, Marquette University, 2016.

[35] A. Fatemi, D. M. Ionel, M. Popescu, and N. A. O. Demerdash, "Design optimization of spoke-type PM motors for formula E racing cars," in *2016 IEEE Energy Conversion Congress and Exposition (ECCE)*, 2016, pp. 1–8.

[36] E. Carraro, N. Bianchi, S. Zhang, and M. Koch, "Permanent magnet volume minimization of spoke type fractional slot synchronous motors," in *2014 IEEE Energy Conversion Congress and Exposition (ECCE)*, 2014, pp. 4180–4187.

[37] Y. Wang, D. M. Ionel, V. Rallabandi, M. Jiang, and S. J. Stretz, "Large-Scale optimization of synchronous reluctance machines using CE-FEA and differential evolution," *IEEE Trans. Ind. Appl.*, vol. 52, pp. 4699–4709, 2016.

[38] Y. Wang, D. M. Ionel, M. Jiang, and S. J. Stretz, "Establishing the relative merits of synchronous reluctance and PM-Assisted technology through systematic design optimization," *IEEE Trans. Ind. Appl.*, vol. 52, pp. 2971–2978, 2016.

CHAPTER **6**

POWER ELECTRONICS AND DRIVE SYSTEMS

6.1 INTRODUCTION

Electrical machines need drive systems to be correctly controlled, if they need to be operated at variable speed. This can be achieved by modulating the energy flow to/from them.

Figure 6.1 depicts a general schematic diagram of a drive system used for electric machines. The following blocks can be recognized from the left to the right: (1) grid, (2) AC-side filter, (3) rectifier, (4) DC-side filter, (5) DC-link, (6) inverter. The power can flow from the AC grid to the machine M, or opposite in case the machine is operated as a generator. A short description of each of them is given in the following.

AC-side filter: The power factor on the grid side must be kept higher than a certain level in order to comply with grid power quality standards. To do this, a line filter bank is used to filter out current harmonics and also conducted electromagnetic interference (EMI).

Rectifier: A rectifier bank is needed to convert the AC power into DC power. Generally, this bank is implemented with semiconductor rectifiers in a bridge configuration. Since they switch on and off abruptly during the grid voltage period, rectifiers normally produce a significant distortion in the AC current waveforms, forcing designers to adopt line filters.

DC-side filter: Very often, and depending on the power rating, the line filter bank can become very expensive or even impractical for reasons like large volume, weight, and therefore an active filtering (AF) solution is adopted, which is called power-factor corrector (PFC) unit. PFCs are power electronic circuits that act as current generators and they are used to shape the line current waveforms.

DC link/capacitor bank: The inverter draws DC power at the switching frequency, which is considerably higher than the fundamental frequency needed to control the machine. For this reason, a buffer capacitor bank is needed on the DC-link connection between the rectifier/PFC bank and the inverter one.

Multiphysics Simulation by Design for Electrical Machines, Power Electronics, and Drives, First Edition.
Marius Rosu, Ping Zhou, Dingsheng Lin, Dan Ionel, Mircea Popescu, Frede Blaabjerg, Vandana Rallabandi, and David Staton.
© 2018 by The Institute of Electrical and Electronics Engineers, Inc. Published 2018 by John Wiley & Sons, Inc.

251

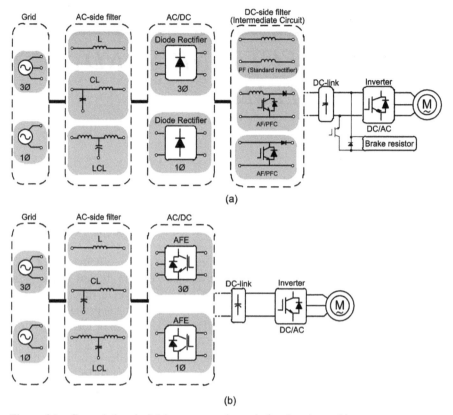

Figure 6.1   General electrical drive system schematic for electric machines.

Inverter: The inverter converts the DC power from the DC link into AC to be fed to the electric machine. As said above, it is operated at the switching frequency, much higher than the fundamental, hence controlled power semiconductor switches are used, which are able to achieve this task. Since the inverter is using high-frequency switching, the AC output power may need to be filtered, in order to avoid high-frequency current flow through the electric machine windings and the bearing's parasitic capacitances.

Referring to Figure 6.1, the power electronics technology is massively used in the rectifier/PFC and the inverter blocks, but modern trends in drive applications are toward using more and more power electronics in other blocks too, including filters and capacitors. Broadly speaking, the reason for that is related to both the reduction in cost and increase in reliability of power electronic components. This allows designers to successfully implement alternative solutions based on frequencies higher than the standard 50/60 Hz, which are beneficial in reducing size, weight, and cost of passives, that is, capacitors and inductors, which might result in large savings [1–6].

From a design perspective, the following challenging aspects can be identified in motor drives applications:

- Losses/thermal design/cooling
- Design margins
- EMI (Electromagnetic Interference)
- Reliability
- Volume/weight

Finally, yet importantly, cost needs to be carefully taken into account when trading off all the above requirements. In fact, the motor drive application business area is extremely wide and a large number of companies share the market worldwide [7].

6.2   POWER ELECTRONIC DEVICES

Power electronic devices operate as a switch, that is, in two possible states: on state and off state. Ideally, in the on state, the voltage across them is zero and an unlimited current can flow through them; in the off state, the current is zero and the voltage drop at their leads can increase unlimitedly. Furthermore, the commutation between one state and the other is normally assumed to be instantaneous. Of course, the above conditions are not reachable in reality, so a number of non-idealities have to be taken into account. Table 6.1 summarizes such non-idealities together with their consequences in real applications. The concepts listed there are key to introduce a correct design approach in power electronic applications and should never be forgotten.

TABLE 6.1   Main non-idealities of power electronic devices

Property	Non-ideality	Consequences in real applications
Voltage drop in on state	The voltage drop is not zero	On-state ("conduction") losses
Current in on state	The current has a maximum	Device choice based on the maximum expected current. Furthermore, overcurrents have to be taken into account
Current at off state	The current is not zero ("off-state leakage current")	Off-state losses
Voltage at off state	The voltage has a maximum ("breakdown voltage")	Device choice based on the maximum expected voltage, including overvoltages
Commutation duration	The commutation duration is not zero	Switching losses (linearly depending on switching frequency)

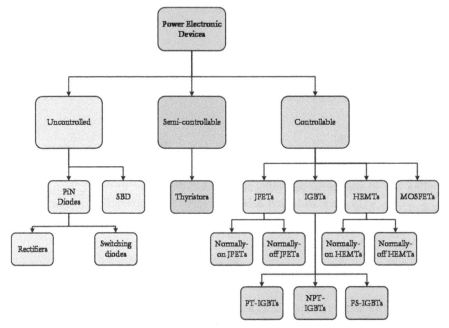

Figure 6.2   Classification of power electronic devices.

A fundamental classification of power electronic devices is based on control-lability, that is, the ability of the user to change the device status from the on state to the off state and vice versa. They can be

a. uncontrolled;

b. semi-controllable devices

c. fully-controllable devices (see Figure 6.2).

Uncontrolled devices are basically diodes, whose status only depends on the current through them; in other words, the user has no controlled way to force a status change. They are: PiN diodes and Schottky-barrier diodes (SBDs). These two categories differ in terms of their internal structure. PiN diodes take their name from the internal structure, which is basically a modified PN junction with an intermediate "intrinsic" layer: P-i-N. PiN diodes are further split into rectifiers and switching diodes, based on the applications they are designed for. In particular, rectifiers are optimized for AC/DC conversion from the grid, hence they operate at 50 Hz or 60 Hz. At such low frequencies, switching losses are negligible and these devices are optimized in terms of on-state losses. Opposite to rectifiers, switching diodes are designed to work at high switching frequency, hence they are optimized in terms of a trade-off between the on-state and switching losses.

SBDs are diodes with a metal-semiconductor junction, the so-called "Schottky junction," named after the physicist W. H. Schottky SBDs perform very well in terms of both on-state and switching losses, but exhibit quite a large off-state current, which in case of silicon limits their practical application to about 100 V. SBDs are becoming largely exploited nowadays thanks to the advent of wide-bandgap semiconductors, in particular silicon carbide, which has allowed to reach 1700 V.

Semi-controllable power electronic devices are basically thyristors, or so-called silicon-controlled rectifiers (SCRs), that is, rectifiers that can be turned on via an appropriate current pulse fed to an auxiliary gate terminal. Anyway, they cannot be turned off in the same way, so the current has to return to zero to switch them back to the off state. Thyristors have paved the way for modern power electronics in the late 1960s, making possible the practical use of switching modulation as a fundamental technique used in adjustable speed drives. Nowadays, though, fully-controllable power electronic devices have outperformed thyristors, taking over switching modulation applications, and confining those latter ones to high-voltage applications.

Former fully-controllable (or controllable) power electronic devices originated from thyristors with gate turn-off thyristors (GTOs) and from bipolar junction transistors (BJTs) with power BJTs. From the 1980s, though, these devices progressively disappeared because of the advent of power junction field-effect transistors (JFETs) and metal-oxide-semiconductor field-effect transistors (MOSFETs), which have been an important breakthrough thanks to the significant reduction in power and complexity of their driver circuit. Nowadays, the scenario of controllable power electronic devices is dominated by MOSFETs, JFETs, insulated-gate bipolar transistors (IGBTs) and high electron-mobility transistors (HEMTs). It is worth to point out, though, that the silicon arena is mostly populated by MOSFETs and IGBTs, whereas the wide-bandgap one is mostly populated by MOSFETs and JFETs, for silicon carbide, and HEMTs for gallium nitride.

The operating principles of the most adopted power electronic devices are given in the following sections.

6.2.1   PiN Diodes

In silicon technology, SBDs are only practical up to 100 V, hence PiN diodes are broadly used in low- and medium-voltage applications and thereby, in motor drives.

Figure 6.3 shows a typical cross section of a PiN diode together with its electrical symbol [8]. The main active region of the device is in the middle and it is the largest one. There, the current and the electric field develop almost in the vertical dimension (1D phenomena). This part mainly consists of three regions: a highly doped P region (P^+), which is connected to the anode contact; a lightly doped middle region (N^-) often indicated as "intrinsic region" or "drift region"; and a highly doped N region (N^+) at the bottom side [9]. Due to this structure, power diodes are also named PiN diodes, the "i" standing for "intrinsic." In discrete devices, the bottom (back) side is electrically and thermally connected to a metal package (not shown in Figure 6.3) which is also the cathode of the device. In power modules, the bottom of the device is connected to a direct-bonded copper (DBC) layer, which provides both electrical isolation and thermal connection. At the active region periphery, there is a termination region whose role is to limit boundary effects on the electric field

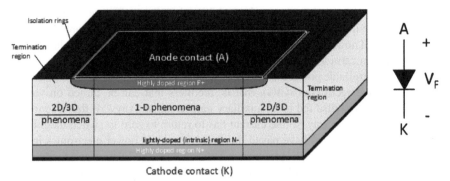

Figure 6.3   Typical structure of a PiN diode (cross section) and its electrical symbol [8].

distribution. Such region does not contribute significantly to the current conduction, but it is very important for the robustness of the diode. In this latter region, 2D and 3D phenomena are dominant. The amount of on-state current that the diode is supposed to carry, defines the cross-sectional area of the chip, whereas the thickness of the N⁻ region (which occupies the main part of the chip's thickness) is dependent on the breakdown voltage that the manufacturer intends to achieve—a feature that is common to almost all the power semiconductor devices. The increase of the intrinsic region's thickness leads to enhanced voltage blocking capability, but at the same time determines the additional ohmic resistance during the on state, the current path becoming longer which degrades the device efficiency. Therefore, a trade-off has to be found during the design process.

On-State Behavior

The real characteristic of a power diode is depicted in Figure 6.4. In opposition to ideal diodes, the on-state current flow does not start when an arbitrarily low positive voltage is applied to the anode contact. In fact, a defined voltage threshold V_T (whose

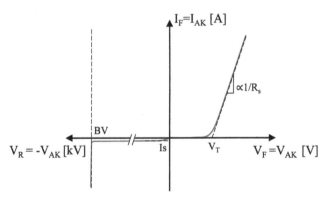

Figure 6.4   Real characteristic of a PiN power diode.

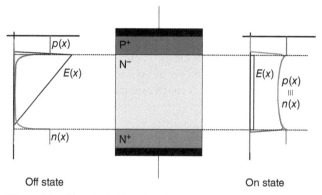

Figure 6.5   Electric field and carrier distributions in off state and on state of a PiN diode [8].

value is 0.7–1.0 V for silicon diodes), has to be reached across the P$^+$/N$^-$ junction in order to overcome the built-in voltage and trigger the injection of carriers, i.e. electrons and holes, into the intrinsic region.

When a larger forward voltage is applied ($V_{AK} > V_T$), the diode is said to be "forward biased". The current flow increases linearly, due to the heavy influence on the voltage drop of the on-state resistance (R_s), which is negligible in ideal diodes. Part of this drop is located across the drift region. Here, when the injection of excess carriers from P$^+$ is considerably high (high forward current), a charge storage phenomenon occurs, resulting in an approximately constant electric field and carrier concentration (typically several orders of magnitude higher than in the off state) across the region, as it can be observed in the right-hand side of Figure 6.5 [8]. Thus, for large current densities, the mobility of the carriers in the intrinsic region is reduced, causing the increase of the voltage drop in an ohmic fashion. Most of the on-state power losses of the device are generated by the current flow through this area. The total forward voltage drop on the diode can be easily approximated as linear, through the following equation [10]:

$$V_F = V_T + R_s I_F \qquad (6.1)$$

Off-State Behavior
When a negative voltage is applied to the anode contact, it will strengthen the built-in voltage and prevent current from flowing through the junction, which is now "reverse-biased." A small current I_s, up to few μA, often referred to as "leakage current" and mainly due to quantum phenomena in the silicon, flows when the diode operates in off state, independently from the reverse voltage level. Leakage current should never be underestimated, since it increases the power losses generation, especially at high temperatures, and, if sufficiently large, can lead to the device failure.

During reverse-biasing, the electric field will expand across the intrinsic region, widening the so-called "depletion region," where the carriers have been pushed away. The function of the N$^-$ region is, in fact, to withstand the high reverse electric field which is usually applied to power semiconductor devices. This electric field assumes a typical triangle-shaped distribution visible on the left-hand side of Figure 6.5. When the reverse voltage level defined as "breakdown voltage" (BV) is outreached, the current flowing through the junction will quickly increase due to a chain mechanism named *impact ionization* and related to the high kinetic energy of the electrons. The operation in the breakdown region has to be strictly avoided since the large amount of dissipated power can rapidly destroy the diode. A rough estimation of the breakdown voltage for silicon is given by [10]

$$BV = \frac{1.3 \times 10^{17}}{N_d},$$ (6.2)

where N_d stands for the doping concentration of the intrinsic region per cubic centimeter. It makes sense then, how high breakdown voltage values are only achievable with lightly doped (and thicker) drift regions. When the depletion region's width (i.e., the base of the yellow triangle in Figure 6.5) at the breakdown voltage does not outreach the N^+ region, the device is named non-punch-through (NPT), whereas, when it spreads beyond the N^-/N^+ junction, the device is called punch-trough (PT). The devices (like PiN diodes, BJTs, or IGBTs) designed with a PT structure usually show lower on-state power losses.

Switching Behavior

All electronic switches require a finite amount of time (usually in the order of magnitude of few microseconds) to turn from the off state to the on state and vice versa. Switching is a critical part of the power device's operation: it is a dissipative process and depends significantly on the features of the circuit in which the device is placed. The nature of an electric load is inductive in most of the applications (e.g., motors or power grids). This means that variations in the load current are not instantaneous, but happen, in fact, with a certain rate di/dt. This influences both the turn-on and the turn-off processes of the diode, as one can observe in Figure 6.6.

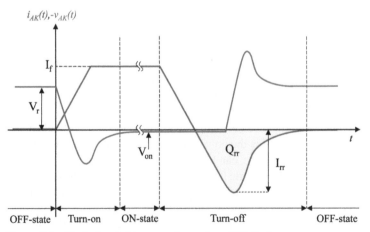

Figure 6.6   Typical switching waveforms for a PiN diode.

The following physical mechanisms occur:

- TURN ON: In order to switch to the on state, the large amount of charges stored in the depletion region has to be removed, as the reverse voltage V_r decreases. Consequently, the current starts rising with finite slope. The voltage across the device will then reach positive values and forward-bias the P^+/N^- junction. Due to the high rate of change of the current, a voltage drop appears on the parasitic (or "stray") inductance which affects the circuit and the device's package. This is added to the voltage waveform as an overshoot. Large overshoots should be avoided, being an additional source of switching power losses and, in some case, dangerous for both the device and the circuit. Eventually, the forward current will reach its steady-state value I_f and the voltage will settle to the on-state drop V_{ON}.

- TURN OFF: During the on state, the intrinsic region becomes flooded with excess carriers, which need to be swept away during the turn-off transition before a reverse voltage can appear again across the junction. As a matter of fact, the current's decay will not stop until the N^- region is emptied out to its original carrier concentration, eventually resulting in a current flowing in the opposite direction, which is referred to as the *reverse recovery current*. When this current reaches its peak value I_{rr}, the blocking capability of the junction is recovered and the reverse voltage starts growing until it stabilizes onto its off-state value V_r, after an overshoot due to the stray inductance. The yellow area in Figure 6.6 accounts for the total charge Q_{rr} that has been swept from the drift region. The reverse recovery is a key phenomenon in many power devices and its duration is linked to other circuit and device parameters. A shorter reverse recovery time is desirable in order to reduce the turn-off delay and power losses. Nevertheless, a necessary design trade-off has to be reached with the voltage blocking capability and the on-state voltage.

6.2.2 MOSFETs

The earliest field effect transistor was patented back in 1925 by J. E. Lilienfeld. Later on, in 1959, the first MOSFET was theorized and invented at Bell Labs [11]. Nowadays, the power MOSFET is the most commonly used power semiconductor controllable switch for low- and medium-power applications, up to several kilowatts [12]. The power MOSFET is widely spread in household appliances, power supplies, and automotive power electronics. The maximum blocking voltage capability is up to 600 V for silicon-based, commercial components. Its operation as power switch does not essentially differ from the low voltage MOSFETs, which are used in digital integrated circuits. Its main features can be summarized as following:

- Fast switching, thus low switching losses
- Low gate driver (control circuit) power consumption
- High-current turn-off capability
- Relative ruggedness to abnormal (fault) conditions
- Relatively large on-state resistance (and conduction losses)

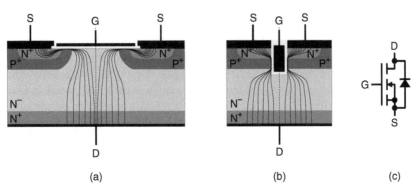

(a)   (b)   (c)

Figure 6.7   Cross section of a horizontal, N-channel power MOSFET's cell (a), cross section of a trench power MOSFET structure (b), and electrical symbol of a power MOSFET (c) [8].

The cross section of a state-of-the-art power MOSFET's cell is represented in Figure 6.7a. It can be observed how a P-i-N layout, identical to the previously mentioned for diodes, is topped by an additional N^+ region and a metal-oxide-semiconductor (MOS) structure, whose metal layer is named *gate* (G). The gate is electrically insulated from the semiconductor by a silicon dioxide (SiO_2) layer and serves as control terminal for the device. The power terminals are the top metallization or *source* (S), and the bottom one, named *drain* (D). This structure is called *enhancement mode*, horizontal *N-channel* MOSFET. The current path between source and drain, or "channel," is basically controlled by the voltage across the gate and source terminals. When the gate terminal is protruding into the semiconductor substrates, like in Figure 6.7b, the structure is called "trench" or vertical *N*-channel MOSFET, with a number of advantages in the achievable power density (the cell width is reduced). Several thousands of these elementary cells (few micrometer wide) are paralleled into a chip to increase the current capability of the device, whose electrical symbol is sketched in Figure 6.7c.

MOSFET Operation: The Field Effect

The structure of the MOSFET prevents it from conducting current if a positive voltage is applied across drain and source, since the P^+/N^- junction is reverse-biased. When a positive voltage is applied between gate and source, positive charge starts accumulating on the metal-oxide interface and consequently, negative charge is attracted in the P^+ doped semiconductor below the oxide surface, not differently from a capacitor. Once the voltage has outreached a sufficiently high value—called the "gate threshold" voltage V_{th}—the layer of free electrons stored under the oxide becomes so thick that a portion of P-type semiconductor inverts its concentration to an N-type, providing a conductive path for the current from the N+ to the N− region. If the drain-to-source voltage is also positive, a current will start flowing. As visible in Figure 6.8 (the *output characteristic* of the device), the higher the gate voltage, the more the current flowing through it, since the channel becomes thicker. For low drain–source voltage bias in the so-called "ohmic" or "linear" region, the current flow is voltage-dependent, because

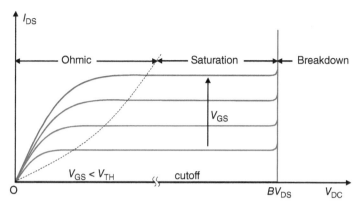

Figure 6.8   Output characteristics of a power MOSFET.

the velocity of the electrons increases with the larger drain bias. When the voltage becomes sufficiently high, the current is no longer voltage-dependent and the characteristic becomes flat for a given gate voltage. The saturated current can be described by the following law [10, 12]:

$$I_D = K(v_{GS} - V_{th})^2 \qquad (6.3)$$

where K is a coefficient depending on the device geometry.

The MOSFET operates in the "saturation" or "active" region until the drain–source breakdown voltage. The working point for a power MOSFET in normal conditions lies in the ohmic region, where the on-state voltage is relatively low and therefore, the conduction power loss is acceptable. The on-state losses of a MOSFET are generated by the current flow through several resistive regions, and are usually larger compared to other devices. The largest contributions to the overall resistance are located in the channel and in the drift region.

When a negative voltage is applied across drain and source, the PiN structure becomes forward-biased and the device starts behaving like a diode. It is important to take into account the presence of this "intrinsic" or "body" diode when designing circuits with power MOSFETs. The diode structure is often optimized by the manufacturer in order to obtain better performances and avoid the use of external diodes, for example, as freewheeling current paths in power converters.

Switching Behavior

The dynamic behavior of the MOSFET is usually much faster than the one of the so-called "bipolar" devices (like diodes or IGBTs). The reason lies in the fact that no excess carrier has to be swept in or out of the drift region, but the transition only depends on the circuit elements and the time necessary to open (or close) the channel. This is defined by the charging (or discharging) rate of the three internal capacitors (C_{GS}, C_{GD}, and C_{DS}) of the MOSFET, visible on the right-hand side of Figure 6.9. The gate–source capacitor C_{GS} is in good approximation constant and its value depends on the geometry of the device. The drain–source and gate–drain capacitor

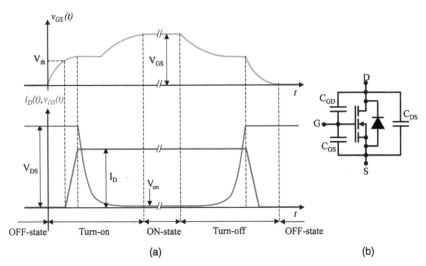

Figure 6.9 Switching waveforms of a power MOSFET (a) and equivalent circuit diagram of a MOSFET during switching (b).

values are instead affected by the drain voltage bias, which shapes the depletion layer in the drift region.

Considering an ideal circuit where the MOSFET is forced to conduct a constant current I_D, the switching waveforms would appear like the ones depicted in Figure 6.9. During turn on, before the gate–source voltage reaches its maximum value V_{GS}, the two input capacitors (C_{GS} and C_{GD}) need to be charged up, while the output capacitor C_{DS} has to be discharged of the full drain–source voltage V_{DS}. The current starts to ramp up as soon as the gate voltage threshold has been outreached until its fixed steady-state value. The opposite mechanisms take place during the turn-off time. The flat regions in the gate voltage waveform, called *Miller plateau*, are due to the discharge (charge) transient of the output capacitor, during which the voltage on the gate terminal is clamped. In a real switching behavior, the high changing rate of the waveforms and the inductive nature of the circuit would cause a current overshoot during turn on and a voltage overshoot during turn off. The circuit controlling the device switching, called *gate driver*, is a crucial element in power converters and has to be designed properly in order to obtain the best performance and reliability.

6.2.3 IGBTs

The IGBT was developed in order to combine the advantages of a MOSFET and the relatively small on-state resistance along with enhanced blocking voltage capability of a bipolar structure [13]. These features have made this device extremely popular in the last 20 years in commercial applications up to hundreds of kilowatts, with outstanding and continuous design improvements. Figure 6.10a shows how these convenient features can be achieved by replacing the N^+ drift region of a MOSFET structure

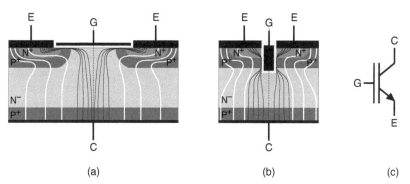

Figure 6.10   Cross section of a horizontal, N-channel IGBT cell (a), cross section of a trench IGBT structure (b), electrical symbol of an IGBT (c) [8].

with a P^+ layer. In this way, a second PN junction is created, which makes the vertical structure of the device behaving like a PNP bipolar junction transistor (BJT). The nomenclature of the power terminals is then changed: the source is named *emitter* (E), while the drain is termed *collector* (C).

IGBT Operation

The physical operation of an IGBT is basically very similar to that of a MOSFET. The additional feature lies in the presence of the N^-/P^+ junction on the collector side. During forward-biasing (positive collector–emitter voltage), when the gate voltage is larger than the threshold, and the current starts flowing through the N-channel, minority carriers are attracted across the collector junction, which is forward-biased, spreading in the drift layer and eventually reaching the emitter P^+ region, since the emitter P^+/N^- junction is weakly reverse-biased. Due to physical phenomena, a current amplification occurs in this process, making the additional current considerably higher than the MOSFET current, and flowing through a less-resistive path (PNP region).

IGBTs cannot have the same reverse conduction capability of a MOSFET, but in fact, it blocks reverse collector–emitter voltage up to the breakdown limit of the collector PN junction.

Switching Behavior

The only significant difference between the IGBT and the MOSFET switching transients is the presence of excess charge stored in the IGBT's drift region. In a manner similar to that of the PiN diode, this charge has to be moved in during turn on, and swept out during turn off, in order to recover the voltage blocking capability of the junctions. Hence, IGBT requires slightly longer switching times, which increases the energy losses. In particular, during turn off, the IGBT exhibits a characteristic current *tailing* phenomenon.

6.2.4 Emerging Semiconductor Technologies

Power semiconductor devices design has been undergoing a continuous improvement process, almost pushing the silicon-manufacturing technology to its physical limits. The maximum possible current and voltage blocking capability has been reached in many cases for single-die chips or *discrete* devices. Nowadays, in order to scale up the available power, many chips are connected in series or paralleled within the same package in the so-called *power modules*. Nevertheless, the investigation on novel semiconductor layouts as well as new materials is still a key point in the field.

Among the most recent developments, it is worth focusing on the diffusion of *wide-bandgap* (WBG) *semiconductors* in the field of power electronic switches [14]. Two materials in particular have shown rather attracting physical features [15] and reached a relative maturity in the manufacturing process: silicon carbide (SiC) and, with still limited applications, gallium nitride (GaN). The WBG-based devices are capable of:

- operations beyond the temperature limits of their silicon counterparts
- enhanced voltage blocking capability with reduced on-state resistance
- increased switching speed, that is, lower switching energy losses

These outstanding features allow the production of faster, smaller, and more efficient devices and power modules, and they are more suitable for harsh environment operations. At the current state of the art, the only factors that hinder the massive diffusion of these devices in the application field are higher prices and the still scarce reliability evaluation.

6.3 CIRCUIT-LEVEL SIMULATION OF DRIVE SYSTEMS

As the complexity and the scale of electric- and electronic-based power systems have been increasing, the need for computer-aided design (CAD) tools of such systems has become rather essential. Modeling and simulating the operation of a circuit before the actual implementation can give essential knowledge about the circuit's behavior and greatly improve the design process, allowing to spot weaknesses and evaluate the performance under different conditions. The simulation is especially needed in those applications, like integrated circuits or printed circuit boards (PCB), where the degree of embedding does not allow a practical prototyping and probing, or in very high power systems, where it would be too expensive to design by trial and error. The simulation process is in fact inexpensive and requires a limited amount of time if managed properly. It relies basically on the numerical solution of the mathematical laws which govern the circuit (mostly Kirchhoff's laws) and its components. Nowadays, most of the modeling and simulation tools integrate a graphical user interface (GUI) with:

- a schematic editor, where one can easily place and connect the circuit components (represented as their electrical symbols) and set their features

- a simulation engine, offering the possibility to choose among a number of solving algorithms and to set the parameters for each of them
- an on-screen output display, where the user can watch the simulation results and waveforms

This makes the evaluation of several "what-if" scenarios fast and practical when changing circuit parameters, using different mathematical models for the components or resetting the simulation's configuration.

A key-point in the simulation process is the choice of model, which will describe each of the components with one or several mathematical equations. While the equations for the standard components (resistors, capacitors, and inductors) are fixed, a large number of models (more or less detailed) for electronic devices are usually available in the software *libraries*. There is also quite often the possibility to define a customized model for a circuit part or to change the model parameters for the existing ones. The models, which are closer to the ideal operation of the device, are usually very simple and do not particularly affect the simulation time, even in large numbers. More detailed models contain a larger set of equations, often complicated, in order to estimate accurately the real behavior of the component. Thus, the solving algorithm requires a relatively long time to converge, especially when many components are included in the simulation. It is fundamental, then, to adjust the degree of accuracy of the simulation, considering the specific goal or application, which is meant to be achieved. In some stages of a design process, it is just worth to obtain an approximate and fast insight on the circuit, rather than an extremely detailed and extensive analysis.

Currently, some of the most popular analog and digital circuit simulation softwares are: OrCAD/PSpice, Saber, Simplorer, MAST, Verilog, VHDL, LTspice, PSIM, PLECS, Multisim, Simulink. The *Simulation Program with Integrated Circuit Emphasis* (SPICE) [16], was developed at the Electronics Research Laboratory of the University of California, Berkeley and initially released in 1973 as a general-purpose, open source analog electronic circuit simulator, coded in FORTRAN. It basically performs a nodal analysis of the circuit, building a *netlist* of the circuit elements, their equations, and interconnections, and translating it into a set of equations to be solved. The equations are mostly nonlinear differential algebraic equations that are usually solved by implicit integration methods, Newton's method, and sparse matrix methods. The typical steps to be followed when performing a circuit simulation are listed in Figure 6.11.

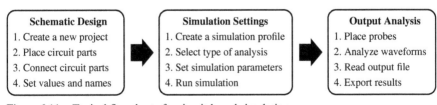

Figure 6.11   Typical flowchart of a circuit-based simulation.

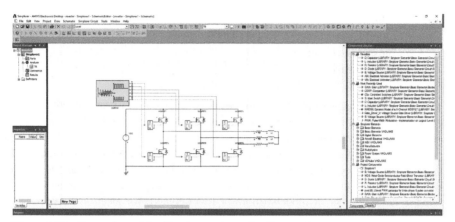

Figure 6.12   Screenshot of the Simplorer schematic editor environment.

6.3.1   ANSYS Simplorer Simulator

Part of the ANSYS Electronics suite, Simplorer is a recently developed analog mixed signal multi-domain circuit simulator. The simulation models can include components from different physical domains, such as thermal, magnetic, or mechanical. Hereby, the software is used exclusively to simulate fully electrical models, built with circuit components in the schematic editor. In the first place, the circuit has to be sketched in the schematic editor, whose interface is shown in Figure 6.12. The circuit parts are chosen from the list available in the libraries and placed on the schematic. A connector tool provides the wiring between the components. At this point, the values and the names for each component can be defined by the user. As a second step, a simulation profile has to be selected. The software is capable to perform the following types of circuit analyses:

- *Transient Simulation*: used to calculate the circuit model behavior in the time domain

- *AC Simulation*: used to calculate the circuit model behavior in the frequency domain. A DC simulation is performed for each operating point and then an AC simulation sweeps a given frequency range.

- *DC Simulation*: used to determine the operation point for circuit models with nonlinear components in the quiescent domain

- *VHDL-AMS Simulation*: a sub-simulator for models described in very high speed integrated circuit hardware description language—analog mixed signal (VHDL-AMS).

Each of these types is characterized by simulation parameters, such as maximum time step or range of values of the analyzed parameters, which have to be properly defined by the user. Display elements may be included on the schematic and the output variables have to be selected before or after the simulation. When the program is executed, the schematic is translated into a netlist and the simulation engine starts

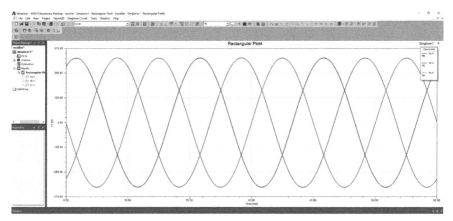

Figure 6.13   Screenshot of the Simplorer simulation results viewer.

running. The solving algorithm is based on a modified nodal approach with numerical integration. The interface allows the user to assist the simulation process and manage convergence issues, in case any should occur. Finally, the output is plotted, like shown in Figure 6.13, in the display elements. The voltage and the current calculated for each component can be viewed in the scope.

6.3.2   Simulation of a Full-bridge Diode Rectifier

In this section, a circuit simulation process in Simplorer will be presented. A three-phase full-bridge *diode rectifier* (Figure 6.14) has been chosen, since it is a standard circuit for AC–DC line-frequency power conversion, used as front end in most of the drive applications. Its function is to convert the AC three-phase voltage supplied by

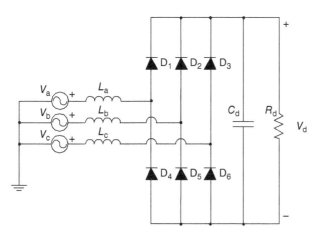

Figure 6.14   Three-phase, full-bridge diode rectifier with finite inductance on the AC side.

the grid into a DC voltage output, using six power diodes. The AC line is modeled as three sinusoidal voltage generators (V_a, V_b, V_c) at line frequency (50 Hz) shifted each other by 120° and connected in series with three inductances (L_a, L_b, L_c), representing the non-ideal behavior of the line. For simplicity, the load connected to the rectifier's output consists of a resistor, while a capacitor acts as a filter. In a real drive system, it would have been an inverter and the drive would act as a constant power load.

In the top group of diodes, only the one with the highest anode potential conducts, while in the bottom group, only the diode with the cathode at the lowest potential is forward-biased. The voltage waveform across the load resistance is expected to assume a characteristic six-pulse shape with an average voltage that is 1.35 times the amplitude of the grid phase-to-phase voltage. The load current always flows through two devices, one from the top group and one from the bottom group and switches among six different paths within a sine wave period.

The effect of the AC-side inductance is to slow down the current commutations between the diodes. When the current changes its path from one to another pair of diodes, the voltage across the diode cannot change, slightly reducing the average output voltage. This is as expected to be observed in the simulation results for this topology. The model chosen for the diode is a standard Simplorer model which features some of the key physical parameters of the real diode. The circuit in Figure 6.15 has been drawn in the schematic editor.

After setting the values for the circuit elements, the simulation profile has to be created, selecting the simulation type. In this case, a transient analysis has been chosen in order to evaluate the voltage and current waveforms in the time domain. A duration of 60 ms and a maximum step size (i.e., the maximum time step which the algorithm can use for the numerical solution) of 1 ms have been set. This should allow getting an insight into more than one period of the line-frequency waveforms in steady state with satisfactory resolution. Once running the simulation, the solution was reached after 0.35 s.

A plot of the voltage waveforms is shown in Figure 6.16 and displays the three voltage sine waves in the mid-point of each leg and the output differential voltage

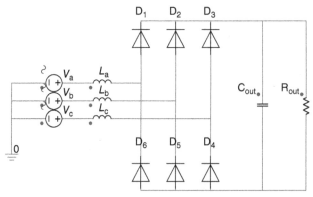

Figure 6.15   View of the three-phase diode rectifier exported from Simplorer schematic editor.

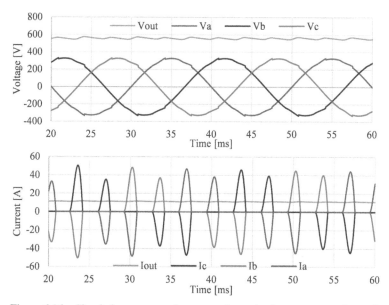

Figure 6.16   Simulation output: voltage waveforms (top), current waveforms (bottom).

across the load. The six-pulse shape, though flattened by the high output capacitance, and the average voltage are in agreement with the theory.

It can be noticed how, in order to obtain a small voltage and current ripple in the load, filtering elements as the output capacitor and the line inductors are introduced, which heavily affect the shape of input current, which appears to be pulsed rather than sinusoidal and phase-shifted. The voltage waveform also appears distorted due to the current commutations between the bridge's legs.

6.3.3   Simulation of a Three-Phase Switching VSI

Another fundamental element for many drive system applications is the DC–AC power conversion stage or *inverter* (see Figure 6.1 too). This is usually connected right after a rectifier, using a DC capacitor bank as interface, in this case it is named as *voltage source inverter* or VSI. A VSI is used in all those applications where the frequency and voltage control of an AC load is required (e.g., in a motor). The control is achieved by regulating the switching of power electronic devices following a pattern, called the *modulation* method. The switching frequency is usually fixed and is much higher than the output waveforms' frequency, which can be variable. In this case, the circuit is called *switch mode*, differently from the line-frequency converters (like the rectifier presented in the previous section). The circuit shown in Figure 6.17 is a typical three-phase, switched-mode VSI based on power MOSFETs. Each couple of series-connected devices (top switch + bottom switch) is called an *inverter leg*.

The control of the switches (i.e., the gate voltage signal) is usually performed by regulating the on-state time in a switching period in order to obtain the desired average

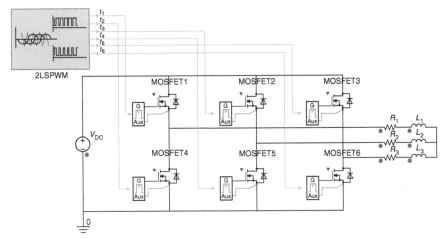

Figure 6.17   View of the three-phase inverter exported from the Simplorer schematic editor.

voltage magnitude and frequency across the load from a constant input voltage V_{DC}. This strategy is called *pulse-width modulation* (PWM). A control signal is compared with a repetitive waveform at the switching frequency, named *carrier waveform*, as shown in Figure 6.18. The shape of the carrier signal (e.g., sawtooth or triangular) can be chosen in order to optimize the degree of usage of the switches and the losses. The level of the control defines the on-state time interval (t_{ON}) for the switch. The ratio of the t_{ON} with the switching period T_S is referred to as *duty cycle* (D) and is proportionally related to the average output voltage V_0 given as

$$V_{\mathrm{o}} = \frac{t_{\mathrm{on}}}{T_{\mathrm{s}}} V_{\mathrm{DC}} = D V_{\mathrm{DC}} \qquad (6.4)$$

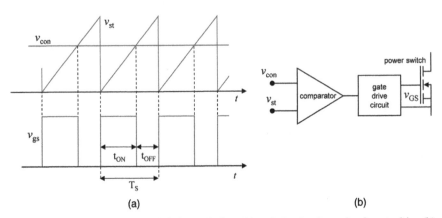

(a)   (b)

Figure 6.18   A pulse-width modulation technique (a) and circuit schematic of a gate drive (b).

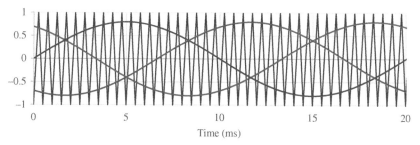

Figure 6.19   Simplorer simulation output for a three-phase sinusoidal PWM.

If the control signal is sinusoidal, then the average voltage in time will also have a sinusoidal pattern with a magnitude proportional to the amplitude of the control sine wave. This is in fact called *sinusoidal PWM* or SPWM. In order to obtain a three-phase output voltage, the control signals should be three sine waves shifted 120° from each other. This provides the gate control for the three top devices. Since the devices on one inverter leg should never be turned on at the same time (to prevent short circuit of the power supply), the driving signals for the bottom switches should be inverted (the ON time of one bottom device coincides with the OFF time of its top one). Also, the power needed to activate the switch is supplied by a *gate drive circuit*, basically an amplifier, which also provides electrical insulation of the delicate components of the control circuit from the power converter. The gate drive is a crucial element for the reliability and controllability of a power converter.

The modulation pattern used in this simulation is the symmetrical three-phase SPWM, whose control signals are plotted in Figure 6.19. A switching frequency of 2 kHz has been chosen, while the sine wave control signals are set to 50 Hz. The simulation output is shown in Figure 6.20. Since the load is heavily inductive, the current rise and fall rate is strongly limited, thus the phase current waveforms in the load look almost sinusoidal. The less inductance in the load (and/or the less switching frequency), the more the output current will show ripple and worsen the harmonic content. The load phase voltage takes the typical five-levels shape: the only possible levels in this three-phase converter topology are in fact 2/3, 1/3, 0, −1/3 and −2/3 times V_{DC}. The negative voltage across the load is achieved by reversing the current in the phases.

6.3.4   Simulation of a PFC

The voltage and the current absorbed by a power converter are actually not even close to ideal sine waves and depending on the nature of the load, they turn out to be phase-shifted. This can be observed in Figure 6.16 for the line rectifier. The phase difference between current and voltage waveform is an important information for the overall operation of a power system and its cosine is defined as *power factor*. When voltage and current are in phase (power factor equals one), their instantaneous product (power) is entirely positive and only *active power* flows towards the load. If the

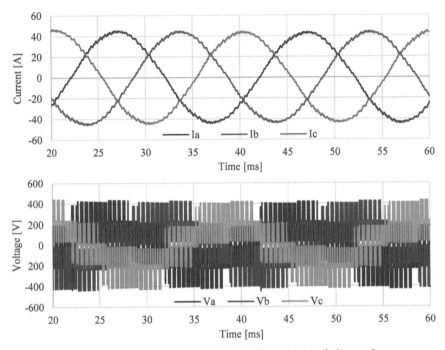

Figure 6.20   Simulation output: phase current waveforms (top) and phase voltage waveforms (bottom).

power factor is less than one, the waveforms are shifted and the instant active power is reduced. A certain amount of power, called *reactive power*, reverses its polarity, flowing back and forth from the generator to the load, causing losses and heating in the whole system and not producing actual workforce. This happens when the loads are either capacitive or inductive, which occurs in most situations.

It is fundamental, when designing a power system, to take into account the effects of power factor and every load is connected to a *power factor corrector* (PFC) circuit to compensate for the reactive power.

Moreover, the power electronic systems and drives usually absorb heavily distorted current and/or voltage (Figure 6.16). This is often expressed by the *total harmonic distortion* (THD) coefficient. A high THD will also cause losses and other instabilities along the line and should be kept as low as possible. PFC circuits have often a THD-correction function as well.

Passive PFCs are simply three-phase capacitor or inductor banks sized to compensate the maximum reactive power absorbed by the load. Active PFCs make use of power electronics to switch the current absorbed by the load in order to obtain a fully controllable power factor and harmonic correction, depending on the operation of the load and the conditions on the line.

A rather simple active PFC circuit, typically used in motor drives, is presented and simulated in this section. The circuit, sketched in Figure 6.21, integrates a

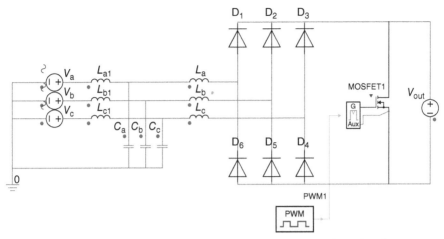

Figure 6.21   View of the boost-type PFC exported from Simplorer schematic editor.

three-phase line rectifier, like the one presented in Section 6.3.2 with a boost-type PFC based on a power electronic switch connected on the bridge output and boost inductors placed on the AC side [17]. The switch, a power MOSFET in this case, is controlled with a simple PWM (like the one in Figure 6.18) pattern. The AC-side inductors are charged during the on state and discharged during the off state. The inductors and the PWM duty cycle are set to generate a *discontinuous conduction mode* (DCM), where the inductors are always fully discharged within a switching period. This causes the phase current to assume a triangular shape with the peaks following a sinusoidal envelope in phase with the voltage, as one can observe in Figure 6.22. The resulting average current is a slightly distorted sine wave in phase with the voltage. The switching frequency harmonic component can be suppressed by a passive LC filter connected on the AC side (Figure 6.23). In this way, as visible in Figure 6.23, the switching frequency component is mainly absorbed by the filter

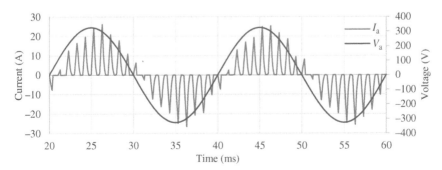

Figure 6.22   Simulation output: filtered phase current and voltage waveforms after the insertion of a PFC.

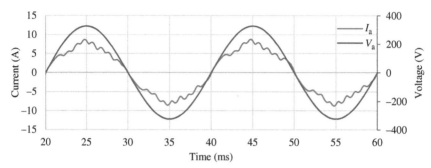

Figure 6.23   Simulation output: phase current and voltage waveforms after the insertion of a PFC.

while the line frequency component flows through the line with some residual ripple (low THD) and in phase with the voltage.

6.4   MULTIPHYSICS DESIGN CHALLENGES

The electrical circuit is not the only part to pay attention to when designing a real converter. In fact, a correct design must also take into account other physical domains, in particular, the thermal one. More in detail, thermal aspects are crucial, as semiconductors require a certain temperature range for correct operations, typically $-40°C$ to $+125°C$, throughout the entire power range [18].

Worst case for thermal design include two conditions:

1. maximum environmental temperature and
2. maximum power losses

In such conditions, semiconductor temperature must never exceed the maximum, stated by the manufacturer. In practice, power losses are seldom constant, but rather vary with time according to a given application. For instance, in inverters they are sinusoidal at double frequency of the output current [19].

Moreover, in order to scale up the power for a given application, a number of devices are connected in parallel into compact, very power-dense modules. Thus, the power losses also increase considerably in a reduced space. In this way, the packaging design process becomes more challenging, with several electrical, thermal, and mechanical phenomena to be taken into account. The basic understanding of a power module structure is crucial for a proper thermal design.

6.4.1   Power Module Structure

Figure 6.24 depicts the typical cross section of a semiconductor IGBT power module. In it, on top of a heatsink (bottom in the picture), there is a copper or other metallic alloy baseplate and a thermal grease is dispensed between them to reduce the interface

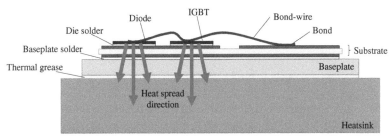

Figure 6.24   Typical cross section of a semiconductor power module.

roughness. The baseplate is soldered to a substrate, generally made up of three layers. A central ceramic tile ensures the electrical insulation between the power devices and the baseplate while preserving a good thermal conductivity. The bottom copper layer is soldered to the baseplate, while the top copper sheet is soldered to the bottom side of the die and its layout is designed according to the circuit topology. The reason of such structure, named *direct bonded copper* (DBC), lies in the low thermal expansion coefficient, which is close to that of silicon and ensure good thermal cycling performances [20, 21]. The different thermal expansion of the layers, in the long run, can determine in fact a mechanical degradation of the bonding and soldering, which can be fatal for the device operation.

The interconnection between the upper side of the dies and the package terminals is provided by bond-wires. Wire bonding is the most used technique to achieve this goal and usually consists of laying aluminum wires with different diameters to form the desired circuit, attaching them to the surface by a combination of pressure, ultrasonic energy, and heat called *thermosonic bonding*. Special machines are capable of performing wire bonding for a module in seconds time scale.

6.4.2   Thermal Modeling

According to semiconductor physics, the semiconductor temperature is also called "junction temperature". Junction temperature is the equivalent thermal network solution illustrated in Figure 6.25, called Cauer-type network [22].

In it, currents represent the heat flux through the circuit elements, and voltages represent the temperatures at certain points. Resistors account for layers' thermal

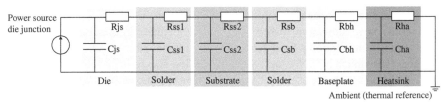

Figure 6.25   Equivalent thermal network for the structure presented in Figure 6.24.

resistances and capacitors account for their thermal capacitances. One can calculate resistances (K/W) based on dimensions and thermal conductivity of a given layer as follows:

$$R_{th} = \frac{x}{A \times k} \tag{6.5}$$

where x is the length of the material (measured on a path parallel to the heat flow), k is the thermal conductivity of the material (W/K·m), and A is the cross-sectional area (perpendicular to the path of the heat flow).

Similarly, capacitances (J/K) can be calculated as follows:

$$C_{th} = V \times \rho \times c_{p'} \tag{6.6}$$

where V is the volume of the material affected by the heat flow, ρ is its mass density, and c_p its specific heat capacity (J/(kg·K))

The thermal network solution represented in Figure 6.25 is relatively simple for static regime, where no heat flux variation occurs. In such a case, capacitors can be taken out and the circuit becomes a trivial resistor network whose solution is

$$R_{th} = R_{js} + R_{ss1} + R_{ss2} + R_{sb} + R_{bh} + R_{ha} \tag{6.7}$$

The total resistance from the above formula must comply with the thermal limits coming from specifications discussed before. In other words, starting from maximum power losses and maximum allowed temperature drop, one can write:

$$R_{th, max} = \frac{T_{j, max} - T_a}{P_{max}} \tag{6.8}$$

where $R_{th,max}$ is the maximum allowed thermal resistance, $T_{j,max}$ is the maximum allowed junction temperature, T_a is the environment temperature, and P_{max} is the maximum amount of power losses. Equations (6.7) and (6.8) can be used to design the thermal structure, in particular base plate and heatsink dimensions and materials.

The above approximation of constant losses is too coarse in real life, where power dissipation has a variable trend and thermal capacitances cannot be neglected. In such a case, even in the simplified network of Figure 6.25, one has to face a complex problem and solution can be seldom calculated in a straightforward way. Moreover, most of the cooling systems require forced air cooling provided by fans or liquid cooling. The presence of cooling fluids introduces even more complexity and needs to be handled with proper multiphysics simulation tools.

Nowadays, simulators based on one of the finest discretization technique employed in computational fluid dynamics called finite volume method (FVM) help designers very much in such a task. Simulators allow specifying arbitrary geometries and custom power profiles against time and solve a specific problem in terms of temperature distribution on fluid flow dependency. The main goal of such approach is to reduce uncertainty and, consequently, expensive design margins.

TABLE 6.2   IGBT three-phase inverter specification

Parameter	Value (unit)
Dc-link voltage	600 V
Output voltage	230 V_{eff}
Output maximum current	40 A_{eff}
Output frequency	50 Hz
Maximum Operating Temperature	100°C
IGBT on-state voltage	1.5 V
IGBT maximum junction temperature	150°C

6.4.3   Thermal Design with ANSYS Icepak

This section presents the thermal management design and simulation for a VSI carried out in ANSYS Icepak [23]. This tool combines advanced solver technology with robust, automatic meshing to perform heat transfer and fluid flow simulation for a wide variety of electronic applications.

Inverters like the one presented in the previous section, need a careful design approach as losses vary continuously during normal operation. A poor thermal management can lead to overheating and thus degrade the reliability of the components.

Table 6.2 reports a sample specification set for an IGBT, three-phase inverter whose Simplorer schematic is depicted in Figure 6.26.

One can calculate worst-case power losses from the previous section, assuming maximum output current according to Table 6.2. A simulation including a detail of current and voltage across a given semiconductor switch is reported in Figure 6.27 while the instantaneous power loss waveform, obtained as a product of them, is plotted in Figure 6.28. The calculated power loss profile can be fed to the thermal simulator to evaluate accurately the resulting junction temperature profile.

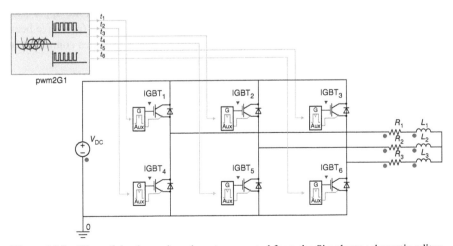

Figure 6.26   View of the three-phase inverter exported from the Simplorer schematic editor.

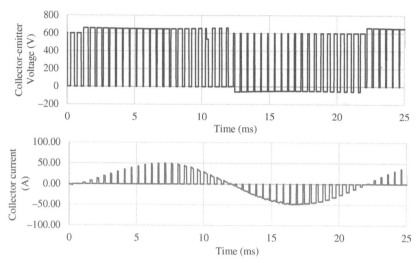

Figure 6.27   Typical voltage and current waveforms across an inverter IGBT switch.

Figure 6.29 illustrates a possible geometry for the inverter shown in Figure 6.26. Three IGBT power modules, each embedding an inverter leg, are placed on an aluminum heatsink. The heatsink geometry is rather typical, with a plate on which the modules are screwed and a number of fins to create a wide thermal exchange surface. A forced air cooling will be provided by a fan placed in the proximity of the heatsink (not visible in the figure). The CAD structure was designed in the ANSYS Space-Claim environment. This allows a direct interface with ANSYS Icepack, simplifying the geometry in order to reduce the complexity of the meshing process.

The structure used in the thermal simulation, however, has been modified to account for the internal layout of the modules. As one can observe in Figure 6.30, showing the simulated structure in the ANSYS Icepak environment, the upper side of the packaging has been removed and the layers showed in Figure 6.24 have been modeled. Each of the switches is, in fact, made up of four IGBT dies connected in

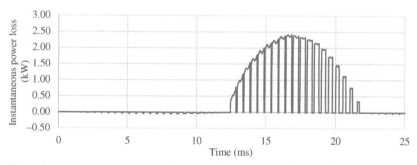

Figure 6.28   Instantaneous power loss waveform across an inverter IGBT switch.

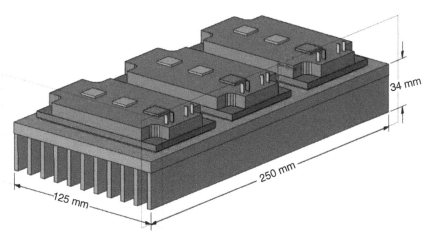

Figure 6.29   ANSYS SpaceClaim CAD structure for the inverter with IGBT power modules.

parallel. Each IGBT has a rated current of 10 A and a maximum operating temperature of 150°C. Thus, the structure has in total 24 devices whose losses profile depends on the modulation pattern in a given time window. The module top and lateral surfaces are assumed as adiabatic in the model, so that the heat exchange takes place only through the baseplate.

A transient simulation has been set up, in order to observe the thermal behavior of the structure for the maximum power losses profile in the operation of the inverter. The ambient temperature has been set to 30°C. The profiles assigned to each device

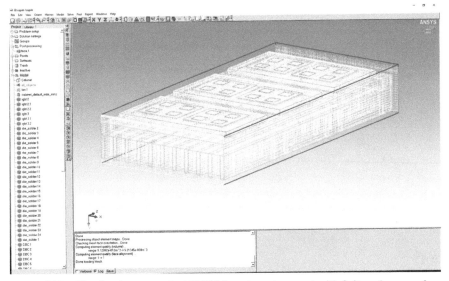

Figure 6.30   Simulated structure in ANSYS Icepak environment with finite volume mesh.

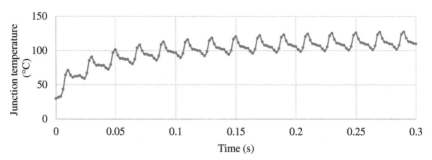

Figure 6.31   Junction temperature trnasient profile.

have been sampled with the same time step of the transient thermal simulation in order to reduce the computational effort. An average of the losses has been calculated for each step. In this way, it was possible to simulate few periods of the 50 Hz waveform, enough to reach the steady-state thermal behavior for the defined geometry. The convergence time can be also controlled by modifying the number of iterations needed to solve for each simulation step. On this particular case study, it was possible to simulate a 0.3 s of thermal transients in about 15 minutes.

The results of the simulation are plotted in Figure 6.31, where one can observe the junction temperature dynamics. The temperature settles well below the maximum allowed for the devices, meaning that the thermal management system has been properly designed. A color map is reported in Figure 6.32, showing the 3D distribution of temperature profile on the inverter structure at the last simulation step ($t = 0.3$ s). It can be noticed how IGBT dies are most critically affected by the heat generation while there is no significant temperature variation in the underlying layers.

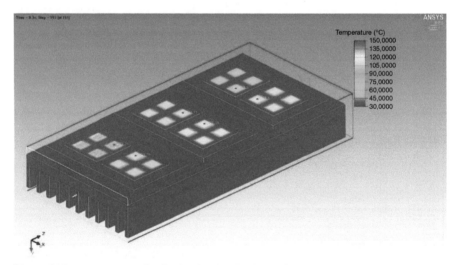

Figure 6.32   Temperature distribution for the simulated IGBT package structure at $t = 0.3$ s.

REFERENCES

[1] W. L. Chen and M. J. Wang, "Design of dynamic voltage restorer and active power filter for wind power systems subject to unbalanced and harmonic distorted grid," in *2016 APEC*, Long Beach, CA, 2016, pp. 3471–3475.

[2] M. C. D. Piazza, A. Ragusa, and G. Vitale, "Design of grid-side electromagnetic interference filters in AC motor drives with motor-side common mode active compensation," *IEEE Trans. Electromagn. Compat.*, vol. 51, no. 3, pp. 673–682, Aug. 2009.

[3] T. M. Parreiras, J. C. G. Justino, A. V. Rocha, and B. J. C. Filho, "True unit power factor active front end for high-capacity belt-conveyor systems," *IEEE Trans. Ind. Appl.*, vol. 52, no. 3, pp. 2737–2746, May 2016.

[4] P. Davari, Y. Yang, F. Zare, and F. Blaabjerg, "A multipulse pattern modulation scheme for harmonic mitigation in three-phase multimotor drives," *IEEE J. Emerg. Sel. Top. Power Electron.*, vol. 4, no. 1, pp. 174–185, Mar. 2016.

[5] P. Davari, F. Zare, and F. Blaabjerg, "Pulse pattern-modulated strategy for harmonic current components reduction in three-phase AC/DC converters," *IEEE Trans. Ind. Appl.*, vol. 52, no. 4, pp. 3182–3192, Jul. 2016.

[6] H. Wang, H. Wang, G. Zhu, and F. Blaabjerg, "Cost assessment of three power decoupling methods in a single-phase power converter with a reliability-oriented design procedure," in *2016 IPEMC-ECCE Asia*, Hefei, China, 2016, pp. 3818–3825.

[7] M. Meza, "Industrial LV motors & drives: A global market update," presented at the Motor and Drive Systems, Orlando, FL, 2014.

[8] F. Iannuzzo, C. Abbate, and G. Busatto, "Instabilities in silicon power devices: A review of failure mechanisms in modern power devices," *IEEE Ind. Electron. Mag.*, vol. 8, no. 3, pp. 28–39, Sep. 2014.

[9] S. Linder, *Power Semiconductors*. CRC Press, 2006.

[10] N. Mohan, T. M. Undeland, and W. P. Robbins, *Power Electronics: Converters, Applications, and Design*, 3rd ed. Hoboken, NJ: John Wiley & Sons, 2003.

[11] W. Shockley, "A unipolar 'Field-Effect' transistor," *Proc. IRE*, vol. 40, no. 11, pp. 1365–1376, Nov. 1952.

[12] B. J. Baliga, Fundamentals of Power Semiconductor Devices. Boston, MA: Springer US, 2008.

[13] B. J. Baliga, M. S. Adler, P. V. Gray, R. P. Love, and N. Zommer, "The insulated gate rectifier (IGR): A new power switching device," in *1982 International Electron Devices Meeting*, 1982, vol. 28, pp. 264–267.

[14] P. G. Neudeck, R. S. Okojie, and L.-Y. Chen, "High-temperature electronics—a role for wide bandgap semiconductors?," *Proc. IEEE*, vol. 90, no. 6, pp. 1065–1076, Jun. 2002.

[15] M. N. Yoder, "Wide bandgap semiconductor materials and devices," *IEEE Trans. Electron Devices*, vol. 43, no. 10, pp. 1633–1636, Oct. 1996.

[16] "The Spice Page" [Online]. Available: http://bwrcs.eecs.berkeley.edu/Classes/IcBook/SPICE/

[17] J. W. Kolar and T. Friedli, "The essence of three-phase PFC rectifier systems," in *2011 INTELEC*, Amsterdam, Netherlands, 2011, pp. 1–27.

[18] "Operating temperature," *Wikipedia*, Jun. 21, 2016.

[19] P. J. P. Perruchoud and P. J. Pinewski, "Power losses for space vector modulation techniques," in *Power Electronics in Transportation*, 1996, pp. 167–173.

[20] "Power electronic substrate," *Wikipedia*, Jun. 30, 2016.

[21] "Efficiency: curamik® Power," [Online]. Available: http://www.curamik.com/index.aspx

[22] G. L. Matthaei, B. M. Schiffman, E. G. Cristal, and L. A. Robinson, "Microwave filters and coupling structures," DTIC Document, 1963.

[23] "ANSYS Icepack: Electronics cooling simulation," [Online]. Available: http://www.ansys.com/products/electronics/ansys-icepack

INDEX

Multiphysics Simulation by Design for Electrical Machines, Power Electronics, and Drives, First Edition.
Marius Rosu, Ping Zhou, Dingsheng Lin, Dan Ionel, Mircea Popescu, Frede Blaabjerg, Vandana Rallabandi,
and David Staton.
© 2018 by The Institute of Electrical and Electronics Engineers, Inc. Published 2018 by John Wiley & Sons, Inc.

IEEE Press Series on Power Engineering

Series Editor: M. E. El-Hawary, Dalhousie University, Halifax, Nova Scotia, Canada

The mission of IEEE Press Series on Power Engineering is to publish leading-edge books that cover the broad spectrum of current and forward-looking technologies in this fast-moving area. The series attracts highly acclaimed authors from industry/academia to provide accessible coverage of current and emerging topics in power engineering and allied fields. Our target audience includes the power engineering professional who is interested in enhancing their knowledge and perspective in their areas of interest.

1. *Principles of Electric Machines with Power Electronic Applications, Second Edition*
M. E. El-Hawary

2. *Pulse Width Modulation for Power Converters: Principles and Practice*
D. Grahame Holmes and Thomas Lipo

3. *Analysis of Electric Machinery and Drive Systems, Second Edition*
Paul C. Krause, Oleg Wasynczuk, and Scott D. Sudhoff

4. *Risk Assessment for Power Systems: Models, Methods, and Applications*
Wenyuan Li

5. *Optimization Principles: Practical Applications to the Operations of Markets of the Electric Power Industry*
Narayan S. Rau

6. *Electric Economics: Regulation and Deregulation*
Geoffrey Rothwell and Tomas Gomez

7. *Electric Power Systems: Analysis and Control*
Fabio Saccomanno

8. *Electrical Insulation for Rotating Machines: Design, Evaluation, Aging, Testing, and Repair, Second Edition*
Greg Stone, Edward A. Boulter, Ian Culbert, and Hussein Dhirani

9. *Signal Processing of Power Quality Disturbances*
Math H. J. Bollen and Irene Y. H. Gu

10. *Instantaneous Power Theory and Applications to Power Conditioning*
Hirofumi Akagi, Edson H. Watanabe, and Mauricio Aredes